Global Resource Depletion, Managed Austerity and the Elements of Hope

Global Resource Depletion, Managed Austerity and the Elements of Hope

André Diederen

Eburon Delft
2010

ISBN 978-90-5972-425-9 (paperback)
ISBN 978-90-5972-426-6 (ebook)

Eburon Academic Publishers
P/O Box 2867
2601 CW Delft
The Netherlands
Tel.: +31 (0)15 2131484 / Fax: +31 (0)15 2146888
info@eburon.nl / www.eburon.nl

Cover design: Studio Hermkens
Typesetting: Textcetera Rotterdam

CONTENTS

Chapter 1 Introduction 1

Chapter 2 Recurring patterns and the laws of nature 5

Probability 6
Thermodynamics 7
Self-organisation: the periodic table of elements 9
Useful patterns to be aware of 11
It's better to be roughly right than precisely wrong 11
Jevon's paradox 12
Over-consumption induced by under-pricing 13
The maximum power principle 14
Liebig's law of the minimum 14
The Pareto principle 15

Chapter 3 The many fallacies of customary convictions 17

"All models are wrong, some are useful" 17
Abundance thinking 22
The meaning of money 22
The refusal to accept physical constraints 25
Grand gesture without constraint 25
The fallacy of cradle to cradle without constraint 27
Technology worship 27
Manipulated convictions and self-deception 31

Chapter 4 Complexity, stability and resilience 35

Introduction 35
Complex adaptive systems, boundaries and leverage points 36
Complexity and the "Law of Diminishing Returns" 39
The Second Law of Thermodynamics revisited 42

Chapter 5 Assessment of limits on resource production rates **45**

Introduction 45
Energy scarcity as central constraint 46
Scarcity of metals 50
Energy scarcity and metals scarcity reinforce each other 55
Scarcity of water and food 57

Chapter 6 Our predicament: overshoot and collapse **63**

Introduction 63
Production rates and reserves of metals 63
Over-consumption and cascading effects 70
Systemic collapse 76
Climate change *78*
Increased prospects for conflict *80*

Chapter 7 Managed austerity and the elements of hope **81**

Introduction 81
Least regret options and adaptability 81
Managed Austerity 83
The Elements of Hope 86
Use less *86*
Longer life *88*
Recycling *89*
Substitution *90*
Product and process design *93*
Buffers *94*
Implementation 95

Chapter 8 Conclusions **101**

Annex A Periodic Table of Elements (cut off after ^{103}Lr) **105**

Annex B Primary Production and Reserves of Metals **107**

CHAPTER 1
INTRODUCTION

Global resource depletion does not refer to the absolute quantity of resources but to the small portion which is within our reach; it's about the attainable quantity at the desired quality per unit of time (say per year). Global resource depletion refers to our predicament that demand for goods exceeds or starts exceeding supply on a global scale. Goods refers to, amongst others, energy, materials, water and food. The potential implications of global resource depletion range from inconveniences, such as limited availability of certain luxuries, to horrific effects, such as mass starvation by water, food and energy shortages and large scale conflicts.

Global resource depletion is not an issue to be taken care of by future generations; it matters *now*. This book offers you the chance to escape from remaining blind, deaf and in denial and gives and explains the generic directions to cope with our predicament, together with the abundance of potential pitfalls. If you care about our collective course, I invite you to read this book, especially if you think and believe that 'the market' will solve our problems or that technology alone will save us.

Is there any reason to be optimistic about our future? Well, for starters, we can fight our problems using our heads instead of fighting each other over the last resources within reach. This book concludes with the Elements of Hope,[1] which have both a physical and a metaphorical meaning. They are the most abundant material elements available on our planet. At the same time, the Elements of Hope are generic

1 A.M. Diederen, *Metal minerals scarcity: A call for managed austerity and the elements of hope*, http://europe.theoildrum.com/node/5239 (2009)

response measures with intrinsic value within a range of credible future scenarios. The future has not been written yet. Real change starts with acceptance. So, as long as there are individuals and groups of people who will actively pursue necessary changes in behaviour and policy once they know and truly understand what's going on, there is reason to remain hopeful.

But don't get carried away by dreaming about some distant perfect future where all our problems will be solved. This book urges you to stay awake and face reality. Wishful thinking is not a strategy. Human creativity and ingenuity by and in itself is not going to help us much without some form of self-restriction, because at the end of the day, all it really does is to facilitate continued resource consumption. Unlimited growth is both unnatural and unwanted, because it ultimately leads to premature death. We simply have to use less to mitigate the worst effects of overshoot and subsequent collapse, to soften the impact. We have to implement some kind of Managed Austerity[1] to withdraw ourselves from our collective rat race of growing and speeding our population and consumption into a dead-end street. It's time we stop hitting the gas pedal, try to find the brakes and steer away from collision course, especially since most of us have enough brain cells to draw a mental street map pointing to alternative directions.

Currently, most of our leaders fail to correctly diagnose our situation to its complete meaning and extent. With a few inspiring exceptions, incomplete and wrong assessments by consultants, lobbyists and policy makers cause our leaders to embrace misplaced optimism and false hopes and provide them and their citizens with a false sense of security. It leads them to pursue various well-intended but suboptimal measures which tend to do more harm than good. As most people do, they think and believe that continued economic growth is normal (wasn't it so for the last many decades?) and that economic shrinkage is the short-lived exception to the 'rule'. But we live on a finite planet. I guess it's easier to visualise our boundaries if we would be one of seven astronauts in a tiny spaceship than being one of seven billion people sprinkled on the outer layer of a giant one where we cannot directly see its boundaries. From the perspective of an astronaut, our planet is just one giant Easter Isle. As long as we stay in collective denial about limits to growth (weren't we warned repeatedly since four decades ago?[2,3]), timely adaptation will not take place. And time is running out.

This book tries to give a comprehensive overview of our predicament, why and how we got here and how we might get out with the least damage. This book offers a credible 'frugal happy ending' if we are willing to be active participants and start pursuing the right directions. Despite the alarming message of this book, its intention is that of an optimist: the thought that a number of readers are open-minded for the facts and analysis provided by this book and are willing to change their convictions and behaviour accordingly. A number of people will see this as a challenge, not as something to be feared. After all, threats and changes always are accompanied by opportunities. The more people become truly aware of the broad picture, the sooner we can start meaningful adaptations and the higher the level of retained safety and prosperity can be. And if nothing else, we'll always have the comfort left of knowing that we have something called human resilience. But we'd better not test our resilience by waiting so long for undisputable evidence to enlighten even the most inflexible souls that reaching such 'total consensus' has become irrelevant. It is far better to prepare. Military planners have the saying "hope for the best, prepare for the rest", and this mindset is highly recommendable in light of our current predicament.
Chapter 2 explains that you don't always have to know exactly all details to get the broad picture of what's going on. Our world and our life are characterised by recurring patterns and by the laws of nature. Important examples are the laws of thermodynamics, the notion of self-organisation and the Pareto principle. But people tend to get carried away by strong notions which work just fine within the boundaries of a certain domain. So, at the same time, we must be aware of the pitfall of seeing patterns where they aren't or applying 'rules' where they are not applicable. This is elaborated in Chapter 3, which discusses amongst others our generally poor insight in probability and the (partly

2 Dennis L. Meadows, *The Limits To Growth - A Report for the Club of Rome project on The Predicament of Mankind* (1972), accessed through the Dutch translation *Rapport van de Club van Rome*, ISBN 9027452466, 1972

3 Donella Meadows, Jorgen Randers and Dennis Meadows, *Limits to Growth – The 30-Year Update*, ISBN 1931498512, 2004

related) misuse of numerical modelling. Other important subjects of Chapter 3 are abundance thinking[4] and associated flawed notions of the meaning of money, 'cradle-to-cradle' without constraint as well as technology worship and self-deception. Next, Chapter 4 discusses complex adaptive systems which establish or re-establish stability using leverage points such as delays and balancing feedback loops. Parts of the earlier chapters are connected to the notion of complexity, so that Chapters 2 through 4 explain our current predicament of global resource depletion and overshoot from an holistic point of view, give meaningful insights and directions how to cope with our predicament and expose the many pitfalls of well intended but potentially harmful sub-optimisations.
Chapter 5 assesses the limits on global resource production rates with a focus on energy and metals scarcity but with the inclusion of the interference with water and food scarcity. Energy scarcity and metals scarcity form a reinforcing feedback loop which has to be combated with balancing feedback loops and other leverage points (from Chapter 4) proposed and discussed in Chapter 7. Chapter 6 argues that our present course of over-consumption is causing cascading effects at an increasing rate. Without strong and timely intervention, this will likely lead to systemic collapse within decades. Climate change aggravates the situation and an outbreak of large scale armed conflicts cannot be ignored as a credible possibility.
Chapter 7 gives a consistent adaptation framework to change our present collective course, taking into account all issues discussed in the preceding chapters. The concepts of the *Elements of Hope* and *Managed Austerity* will be explained. The Elements of Hope have both a metaphorical meaning and a literal meaning of applying living nature's building blocks and include recycling, substitution and product or process design. But without some form of restriction or control, at the end of the day, these measures will just facilitate continued consumption growth. So in addition to the Elements of Hope we need to implement Managed Austerity, which leads to both lower production, consumption and conversion rates and lower absolute levels of resource consumption. Finally, Chapter 8 gives the overall conclusions.

4 John Michael Greer, *The logic of abundance*, March 24, 2010, http://thearchdruidreport.blogspot.com/

CHAPTER 2

RECURRING PATTERNS AND THE LAWS OF NATURE

The essence of almost everything lies in the repetition, combination and variation of preferential patterns. Learning is recognising patterns. We can recognise the repetition, combination and variation of preferential patterns in divergent things like vibrations, waves, rhythms, music, dance, crystals and the fabric of plants and animals and life itself. Preferential patterns involve variations around some state of equilibrium; this balance in turn can vary towards another state, and so on. One way of looking at atoms is seeing them as little particles that are dancing in perpetual motion, attracting each other when they are a little bit apart but repelling each other when they get too close. Similar patterns occur on a cosmic scale. The 'dance' can be obscured to the naked eye by the length scale (very small or very large) or by the timescale (very small or very large) involved. And as we can learn amongst others from gravity, it takes dance partners or counterpoints to make patterns to start with. In other domains people use terms such as 'ying and yang', and from the behavioural sciences we know the observation that the domination of certain aspects of behaviour (e.g., rationality) usually leads to a deficit (or even neglect) of 'opposite' aspects of behaviour (e.g., empathy). As with most things in life, it's a matter of both absolutes and relatives simultaneously. Time plays a crucial role, often in a nonlinear way which is hard to grasp using the rational part of our mind. Think, for instance, about a piece of music: the enjoyment of music lies not only in the notes being played right now but also in the knowledge of what is coming next. It lies within the 'dynamics' (directions plus rates of change), so to speak. Similarly, the rejoice beforehand or the journey itself can be much more gratifying than reaching the destination itself, as probably most of us already have experienced. Instead of ending patterns, it's usually much more gratifying to continue them, to vary them into another direction or to pursue another pattern (another journey). And our current predicament ('bad vibes'), explained further in this book, urges us to pursue another journey, to change pattern.

PROBABILITY

The world is governed by laws relating to probability, not to certainty. Probability is intrinsic to nature, as follows from the observation that practical physics is not the physics of the individual particle but the physics of the collections of particles ruled by probability. We cannot predict with exact precision both the position and the impulse of a particle (e.g., an electron); we can only predict the probability of these parameters having certain combined values (Heisenberg's uncertainty principle). Of course this particular uncertainty is being cancelled out with larger numbers of particles, but uncertainty resurfaces at larges scales. For instance, it is easier to assess the probability of a large collection of particles having certain collective characteristics than doing the same for only a few particles. Under a microscope, molecules in a fluid or in a gas (e.g., cigarette smoke particles) show seemingly random motion whereas to the naked eye the substance in question seems to be more or less in rest as a whole. Taking a big step to the level of weather events: we cannot predict the exact order of weather events around our house in some small radius (say 500 metres), but we can estimate the probability of having sunshine or rain or whatever in the coming days within the wide surroundings (say 50 kilometres). For any probability relating to matters involving human behaviour (like economics), predictions and 'rules' are vastly trickier. Humans tend to have the nasty habit of exhibiting *free will*, thereby stubbornly refusing to fit our nice rational models. Humans also tend to have a poor feel for probability and this is another partial explanation why "all models are wrong, some are useful" as will be discussed in Chapter 3.

It is impossible to predict the exact order of future events, but some useful predictions can be made with regard to the *probability* of future events. You really cannot predict what would happen in a given circumstance, you can only predict the *probability* of different events. People search for certainty but ironically there is no certainty. You cannot avoid surprises. This is the reason why any 'plan' for the future should be *adaptive*, as will be discussed in Chapter 7.

THERMODYNAMICS

The law of gravity is quite obvious: what goes up must fall down (with absence of propulsion). But the second law of thermodynamics is less obvious and is being violated frequently in proposals and schemes to enable continued growth in energy and materials consumption. That second law means, amongst other things, that energy conversions will always be accompanied by losses. It takes energy to concentrate energy and if the rate and scale of our global energy consumption continue to be kept well above ambient levels, we'll soon hit physical limits. Let me explain.

The amount of work you get out of a given energy source depends not only on the amount of energy it contains but also on the difference in energy concentration between the energy source and the ambient environment. This is generally poorly understood by undifferentiated proponents of 'renewable' energy sources such as solar: they emphasise the total amount of energy ("the sun gives us enough energy in half an hour to run our industrial civilisation for one year") whilst downplaying or even neglecting the fact that solar energy has a poor concentration compared with fossil fuels. Enter the concept of 'exergy'. The exergy is the amount of useful work that the energy source is able to perform. So, besides quantity, it is also quality that counts. And further on in this Chapter (see 'maximum power principle'), we'll discuss that higher conversion rates lead to higher exergy losses. The total quantity of solar power available from the sun may be enormous, but the exergy is so low that direct use of solar power can only propel a covered-bike-like one-person spartan vehicle if that vehicle has to cover large distances at high speeds. And if we try to boost the amount of energy and the attainable power levels by using indirect solar power (or any other means of renewable energy) by using large stacks of batteries, we will have left the affordability zone for mass application at warp speed before we have matched the performance and comforts of a high exergy fossil fuel-driven car. We do have the technology, but the inconvenient question is: at what cost? Without fossil fuels, the only real (affordable) solution here is to accept a big trade-off in speed, range and energy-consuming luxuries onboard. Or we have to abandon personal motorised transport on the current scale in future. Of course solar energy is vital for our future energy needs, but some of the current ideas and developments may do more harm than good with large scale implementation within the holistic approach of our current overshoot. This will be elaborated in Chapter 6 after the introduction and discussion of important facets like complexity (Chapter 4) and materials scarcity (Chapter 5).

Because renewable energy sources have such low concentrations compared with fossil fuels, it takes more materials (like concrete and metals) to harvest and convert them into a useful form like electricity. And as we'll see in Chapter 5, the issue of having to take both quantity and quality into account resurfaces with metals scarcity. The lower the ore grade and the accessibility of the mineral from which the metal has to be extracted, the more exergy you need as input to concentrate the metal. So we need lots of energy in the desired form (like diesel fuel and electricity) to concentrate and produce the metals required to harvest and convert renewable energy. And unlike the situation with fossil fuels, most of the desired and required infrastructure for renewable energy has yet to be built. In short, one of the dilemmas we're facing is the fact that you need exceedingly more energy (through the required materials) to harvest and convert energy.

Another important concept is 'entropy', or the level of disorder in a system. If we have a little box with a layer of fine salt grains and another layer of pepper powder on top of that, the entropy of this 'system' is very low. The system is said to be very ordered. If we shake the box, the salt and pepper will be mixed very soon, the system now is quite disorderly and has a high entropy. You can image that it takes almost no effort to mix the salt and pepper. If you have the proper dimensions of salt grains and pepper powder, you also can let the mixing take place much more slowly 'by itself' by small vibrations (from people walking by or slamming doors and so on) by leaving the box in a kitchen cupboard. Now try to imagine the effort it would take you to reverse the condition of the box and separate all the salt from all the pepper. You now have a taste of the non-reversible nature of most processes in nature, going from a less likely state (perfect separation of salt and pepper, low entropy) towards a more likely state (mixed, high entropy) if left to their own. So a system left by its own strives for maximum entropy. This is being counteracted by the simultaneous strive of a system left by its own to a combination of a minimal energy content and the most stable state. This is being combined as the strive for a minimal free energy content and this is a basic driver of *self-organisation* as can be witnessed in nature, e.g., crystal formation in saturated solutions (making salt out of seawater by evaporating the water). In Chapter 4, we'll see that

self-organisation also emerges at the macro level of complex adaptive systems and is directly linked to system resilience.
Energy can neither be created nor destroyed (the first law of thermodynamics), and left to itself, it always flows from higher concentrations to lower concentrations (the second law of thermodynamics), meaning that you can't concentrate energy without using energy to do it. Then how can we humans constantly use concentrations of energy above ambient levels? The answer is that we take energy from the flow of sunlight through the system of the earth in various ways, as food and also as stored sunlight in the form of fossil fuels. We've been spoiled with our abundance of fossil fuels, the result of many millions of years of concentrating solar energy into chemical bonds which we are now rapidly depleting within a mere few centuries. A basic implication is that we have to be frugal on using the remaining high exergy sources of fossil fuels by making a real and sound distinction between luxuries or 'nice to have' (hint: many) on one hand and necessities or 'need to have' (hint: few) on the other hand. This will be further discussed in Chapter 7 with the concept of Managed Austerity.

SELF-ORGANISATION: THE PERIODIC TABLE OF ELEMENTS

As you probably know, our physical world is built from tiny, indivisible particles called atoms. All atoms are composed of electrons, protons and neutrons, which are the three elementary building blocks of atoms. You can imagine that the number of possible arrangements of these three elementary building blocks of atoms is virtually infinite. However, the self-organisation in nature manifests itself here by restricting the number of 'stable' arrangements of electrons, protons and neutrons in the form of atoms to a number of roughly 100 *elements*, around 80 of them being metals. Each element consists of a nucleus (protons and neutrons) surrounded by a 'cloud' of electrons. The numbers of electrons, protons and neutrons in an element are more or less proportional to each other. Electrons cannot occupy any orbit around the nucleus; they can occupy only a discrete number of orbits because the wavelength of an electron has to fit the circumference of the orbit an integer number of times. Electrons can jump from orbit by absorbing or transmitting discrete quantities of energy, especially the electrons in the outer orbit, so the (outer) electrons largely determine the chemical properties of the element.

Annex A gives the periodic table of elements (in a simplified form) which organises the elements into rows and columns. A thorough, inspiring and beautifully illustrated overview of the complete periodic table of elements is given by Theodore Gray[5]. If you haven't bought this book yet, you can look up the elements on the various web pages dedicated to the periodic table of elements. The lower the row, the more electrons, protons and neutrons the element contains and the heavier it gets. Elements with equal numbers of electrons in their outer orbits are organised in columns. So elements in the same column have similar chemical properties. To a lesser extent, this also is often the case for elements in adjacent columns. So a rule of thumb is that elements close to each other in the periodic table of elements can be each others substitute for their main applications and are also often found in the same mineral deposits. Of course this is just a rough approximation, there are plenty of exceptions to this rule of thumb, but it is a good starting point nevertheless. You can immediately recognise the platinum group metals as a block of six (^{44}Ru, ^{45}Rh, ^{46}PD, ^{76}Os, ^{77}Ir, ^{78}Pt). You also can see why copper (^{29}Cu), silver (^{47}Ag) and gold (^{79}Au) often are encountered in the same mineral deposits. And the lanthanides (^{57}La through ^{71}Lu) were once even thought to be one and the same element. Their chemical properties are so similar that it took us until the twentieth century to be able to separate them.

The elements in Annex A within the thickened section are metals, with the exception of silicon (^{14}Si), which is a metalloid. The metal technetium (^{43}Tc) is also exceptional, because it is synthetic, it is a short-lived radioactive element used for medical diagnostics. LA stands for the lanthanides, which are the so-called rare earth metals, and AC for the actinides which are all radioactive. Some include the elements scandium (^{21}Sc) and yttrium (^{39}Y) among the rare earth metals because they lie in the same column (3a) as the lanthanides. The elements in column 1a, with the exception of hydrogen (^{1}H), are the alkali metals, the next column (2a) are the alkali earth metals. The elements of columns 3a through 2b are the so-called transition metals. The remaining elements within the thickened section (except the metalloid silicon)

---

5 Theodore Gray, *The Elements – A Visual Exploration of Every Known Atom in the Universe*, ISBN 9781579128142, 2009

are the so-called ordinary metals. In the most right column, we have the noble gases; column 7b contains the halogens, and the remaining elements of columns 3b through 6b outside the thickened section are the non-metals. But other definitions exist. Gray[5] reckons the elements ^{5}B, ^{14}Si, ^{32}Ge, ^{33}As, ^{51}Sb, ^{52}Te and ^{84}Po to the metalloids.

The periodic table of elements plays a key role in assessing our predicament with regard to global resource depletion as all main scarcities are related to scarcity of fossil fuels, metals, water and food, all of which, of course, are composed of these elements. At the same time, the periodic table of elements is quite useful in visualising the holistic solution direction of preferred application of the most *abundant* elements, as will be discussed in Chapter 7.

USEFUL PATTERNS TO BE AWARE OF

The human brain is hardwired to recognise patterns, sometimes within a split second. This proved to be quite convenient in ancient and dangerous times when we couldn't afford to mistake a herd animal from a predator (it doesn't matter in the safety of a zoo these days). This pattern recognition forms the basis of learning. Of course, the drawback of our brain being hardwired for pattern recognition is the danger of seeing patterns where they aren't. This is especially true if we feel out of control. Our brains can imagine patterns that don't really exist. People want to think that at least someone is in control, even if that isn't the case. So it makes sense that conspiracy theories become more popular in bad or stressful times. But most of the time, no conspiracy is going on. The 'truth' (as if there would be only one) is not so much being covered up but is largely being lost in the excessive noise of high entropy 'information'. More often than not, it is quite probable that the people 'in control' lack a coherent vision about what's going on, especially if very complex issues are involved. So the first pattern to be aware of is the pattern that is not really there.

Of course, some patterns are really there.

IT'S BETTER TO BE ROUGHLY RIGHT THAN PRECISELY WRONG

Except, of course, while performing brain surgery or designing a nuclear power plant.

We see the pattern of knowing the trees and not seeing the forest occurring more or less consistently throughout various domains, and obviously the domain of global resource depletion is no exception. People in the mining industry, for instance, who see certain by-product metals being discarded as a waste stream are inclined to dismiss the notion of metals scarcity and see it as alarmist. But the notion of depletion of resources within 'easy' reach cannot be treated as an exact science for which it only takes a single exceptional observation to refute an hypothesis. Global resource depletion is governed by a complex set of factors including not only physical science but also geopolitics and geo-economics and hence human behaviour.

The problem one often encounters when arguing with entrenched 'subject matter experts' is their lack of a broad view and their inability to acknowledge the differentiated view that both the viewpoint of the expert in question as well as the viewpoint of one proposing the 'broader view' can be correct to certain degrees simultaneously. But, disturbingly, many experts have a digital view on this: "you are either right or wrong", resembling the oversimplification of politics into 'left' and 'right'. And when narrow-minded experts (they can be engineers, economists or whatever) realise they can't refute your message, they either change the subject or try to discredit you by lack of solid arguments. Again, we see this in both science and politics, and this does not come as a surprise of course since in both cases human behaviour is involved. Fortunately, some people combine the abilities to see the trees and the forests or achieve this as a group by open-minded discussion and cooperation among people with different views.

JEVON'S PARADOX

A recurring pattern is the rebound effect, also referred to as Jevon's paradox. Jevon's paradox is the proposition that technological progress that increases the efficiency with which a resource is used tends to increase (rather than decrease) the rate of consumption of that resource. So, technological progress on its own (without 'control') will only accelerate the depletion of resources. That is one of the reasons for the need for Managed Austerity, as will be discussed in Chapter 7.

There are plenty of examples of the rebound effect. If we make cars that get better mileage, we drive more per car and we make and buy more cars. If we make the construction of the car lighter, we add all

kinds of other features. If we make large fuel efficient passenger jets, air travel becomes a mass consumer good.
The rebound effect also is linked to the general observation from the perspective of systems dynamics (see Chapter 4) that we often identify the correct levers for change but push them in the wrong direction. If the problem is too much plastic garbage and we build incinerators to burn it and make electricity in the process, the 'solution' of producing electricity efficiently from plastic garbage requires sustained or even elevated production rates of plastic garbage, not less garbage. A better solution may be to intervene on the demand side by pricing or taxation, but this has to be accompanied with several other solution directions to prevent unintended consequences. The pricing or taxation intervention is, as almost all measures, no cure for all and may partly have to give precedence to rationing.

OVER-CONSUMPTION INDUCED BY UNDER-PRICING
A pattern which to some extent is linked to Jevon's paradox is the observation that any good or service which is under-priced will be over-consumed. A notorious example is the waste of huge amounts of 'free' river and ground water by lack of incentives to use that water effectively and efficiently, resulting in the growing depletion of aquifers worldwide and resulting in rivers not reaching the shoreline anymore during summer (or even longer). Another example is the 'free' use of our environment as a giant waste dump for toxic substances (gaseous, fluid and solid), plastics (ocean pollution), greenhouse gases and all kinds of waste. By this type of under-pricing, we are in fact borrowing from our future generations who will have to pay somehow for the unpaid part of the bill. Like Jevon's paradox, solutions to cope with over-consumption induced by under-pricing will be discussed in Chapter 7. The simplest solution is preventing under-pricing, meaning correct pricing or taxation (preferably without increasing overall taxes). We may have to revert to rationing in some cases (requiring buffers), both to have a timely effect and to have a distribution mechanism which is more equitable.

THE MAXIMUM POWER PRINCIPLE

The maximum power principle means that a beneficial compromise exists between speed and efficiency of energy conversions, a compromise which gives the maximum power of useful energy conversion at intermediate efficiencies. This is directly linked to the second law of thermodynamics: only an infinitely slow process would approach 100% efficiency, whereas a fast process usually is accompanied by a high conversion rate and a high energy concentration difference and goes hand-in-hand with a high conversion loss. For this reason and for reasons of materials scarcity (see Chapter 5), it might be wise to focus more on using harvested low-exergy solar energy in the form of low-exergy heat at the expense of a moderate conversion loss than focus on converting most of the harvested energy into high-exergy electricity at the expense of a high conversion loss. Solar panels which produce both electricity and hot water have been applied for decades now; they exploit more of the exergy of solar radiation than panels that only produce electricity.

We see the maximum power principle on various levels in nature. It doesn't mean that high rates can't be useful. The millennia-old 'system' of lions as predators of herd animals has been quite successful. But of course, lions don't run around all day at top speed. Most of the time, they have a low 'rate' to enable a short burst at a high 'rate'.

LIEBIG'S LAW OF THE MINIMUM

Liebig's law of the minimum states that production is limited by the scarcest necessary resource. This law originates from biology but is much more widely applicable. In biology, the case is clear: if plant growth requires sunlight, nutrients (like phosphorous, nitrogen and potassium) and water, then a lack of water can't be compensated by increasing the amount of sunlight or nutrients. The same is valid for the other required inputs for plant growth. A direct consequence of this easy to understand concept is that we need not be too concerned for the coming decades by envisioned shortages (somewhere from 2050 onwards) of abundant and cheap phosphate rock-based fertilisers, because we already suffer now from water shortages worldwide. As the saying goes: key to long-term survival is short-term survival. The threat of global food shortages is very real and encompasses various factors, including the need for phosphate fertilisers, fossil fuels (for amongst others nitrogen fertilisers, pesticides, herbicides, fuel for tractors), soil erosion and climate change. But the most stressing and limiting factor for global food production is water scarcity, which also is linked to energy scarcity (see Chapter 6).

THE PARETO PRINCIPLE

The Pareto principle is the principle of unequal distribution and is covered extensively in Richard Koch's book.[6] It is an observed 'universal' pattern of an imbalance between effort and reward, between cause and effect, between inputs and outputs. The Pareto principle is about the difference between those few approaches that do things very well and the vast mass of mediocrity. It is linked to our fallacy (see Chapter 3) of thinking in terms of averages and overall performance. It is about nonlinearity, even the nonlinearity of the distribution of fossil fuels and rich mineral ore deposits across the globe (Chapter 5). It touches on the concept of complex systems theory (see Chapter 4) which covers leverage points: due to nonlinearities, small changes can have huge effects and this can act against us (Chapter 6) and also can be put to good use (Chapter 7). This is linked directly to *hope,* which I briefly touched on in the introduction (Chapter 1) of this book: the balance of circumstances can be shifted in a major way by a minor action, or can be shifted in a major way by a major action by only a few people. Major concepts of the Pareto principle are selectivity and simplicity and these are important notions for an adaptation framework (Chapter 7). Generally, we are pursuing too many directions and doing too many things for too many purposes. Only a few things are really important; most are not. We should be selective, not exhaustive. The more we have to choose from, the less we are inclined to make a choice. And too much information can be harmful: information can become very confusing without a proper goal in view. And as Koch puts it: "go for the most simple 20%, because the simplest structures and process flows are the most attractive and have the lowest cost"[6]. We'll revisit the Pareto principle and Koch's effective description of it in Chapter 4 on complexity.

6 Richard Koch: *The 80/20 Principle*, ISBN 1857881680, 1998

CHAPTER 3
THE MANY FALLACIES OF CUSTOMARY CONVICTIONS

Once you (or others) believe a mental map, it's very hard to challenge it. Beware of your (or others') confident convictions.

"ALL MODELS ARE WRONG, SOME ARE USEFUL"

People often confuse the descriptive quality of models with their perceived predictive power. Most models are descriptive, not predictive. Models are based upon correlation, but even if the applied correlation is correct, correlation reports the past but it doesn't predict the future. Models are always abstractions of (real or perceived) reality and therefore should always be applied using common sense.

Let me start this part on the fallacies related to modelling with our generally poor insight in probability. You should easily be able to refute the proposition that you have a 50% chance of being killed if you try to cross a busy road: you either will be killed or you won't. But let me explain the pitfall of 'knowing' probabilities using a less obvious example. Most people aged 40+ are familiar with the 1970s-style TV quiz show on which you can choose your prize by picking one out of three chambers with closed curtains. Only one chamber has the big prize, for instance a car. The chance that you pick the correct chamber is one in three. Let's suppose that after you have made your choice of the 'winning' chamber, you have to open one of the other two chambers and the one you choose appears to be empty or gives you a one-year supply of cat food. The quizmaster now offers to let you change your original pick to the other closed chamber. What do you decide? Most people think it doesn't matter because the odds of the car being in your original pick and in the other chamber are the same, right? No, wrong!

The chance of the car being in your original pick is still 1 in 3, but the chance of the car being in the other closed chamber has risen to 2 in 3. That is because at the start (with all 3 chambers closed) the chance of the car being in one of the chambers not picked by you are 2 in 3, divided as 1 in 3 for the one and 1 in 3 for the other chamber. As soon as has been revealed that one of those 2 chambers does not contain the car, the chance of the opened chamber having the car decreases to 0 in 3 and for the still closed chamber it rises to 2 in 3, because the chance of the car being in the chambers other than the one you originally picked will keep totalling 2 in 3.

It gets even clearer using the newer example of what in The Netherlands gets aired as the "hunt for millions" show, in which amounts varying from 1 Eurocent to 5 million Euros are divided among 26 closed cases. The candidate may pick one specific case at the beginning, so the odds of him/her having chosen the case containing the 5 million Euros is 1 in 26. The chance of the 5 million Euros being somewhere in the other cases is 25 in 26. A big screen shows all the money prizes, 26 in total (one per case). The candidate has to pick the cases to be opened, after which the corresponding prizes are revealed and withdrawn from the game and from the screen. After a number of case openings, the candidate is offered an amount of money to quit the game. The money offered depends on the prizes left in the game. Now suppose that someone is lucky enough and brave (or stupid) enough to end up with only his own case and a single other case left in the game while the 5 million Euro case is still in the game. The chance that the candidate's case contains the 5 million Euros is 1 in 2 because only two cases are left in the game, right? No, wrong! The chance of the main prize being in the candidate's case is still 1 in 26, whereas the chance of the main prize being in the only other case left has risen to 25 in 26. That's because right from the beginning, the chance of the 5 million Euros being somewhere in the other cases remains 25 in 26. But as the game evolves and information becomes available about opened cases not containing the 5 million Euros, the chance of the other cases containing the main prize increases. Precisely because of this counterintuitive probability distribution towards the end of the game, most people tend to grossly overestimate their chances of having a big prize in their own case and sometimes stubbornly continue gambling until they end up with an amount of money well below the maximum amount of money they were being offered earlier during the game to quit. If an impractically high number of shows could be organised and played, we would have hundreds of peculiar shows ending up with only two cases, including the one with 5 million Euros, but in only 1 in 26 of these peculiar shows will the 5 million Euros appear to be in the case originally chosen by the candidate.

We see a somewhat similar failure to grasp the true meaning of probability by the use of the normal distribution function (Gaussian function, bell curve). As I stated in Chapter 2, patterns are essential to life, to nature, to understanding and learning. But the pitfall is to see and apply certain patterns where they are not applicable but nevertheless seem to be so. The normal distribution assumes symmetry and underestimates the probability of large deviations from the mean or 'normal' value. Nassim Nicholas Taleb[7] gives a nice example I want to use here to illustrate the danger of assuming 'symmetry'. He "asks" streetwise guy 'Fat Tony' and the serious academic 'Dr. John' what they think of the odds of a coin giving heads or tails the next flip, assuming that the coin has an equal probability of coming up heads or tails when flipped and given that the previous 99 flips delivered heads each time. Dr. John replies that the odds are 50% either way because a fair coin gives that chance each time independently from previous flips. Fat Tony replies no more than 1 percent, because he thinks the coin has to be loaded, it cannot be a fair game to his opinion given the 99 previous results. Who do you think is more likely to be correct? The danger of underestimating the probability of large deviations from the 'normal' value can be illustrated by the inability of even the most sophisticated numerical models running on the mainframes of large financial institutions to correctly predict the latest financial crisis. People from within the financial sector (or affiliated with it) who correctly predicted the latest financial crisis of course used computers as well, but they largely relied upon their own skills, experience, knowledge and sound judgement. Computers for them are a valuable tool, not some sacred oracle giving predictions to be followed blindly.

The 'misuse' of (numerical) modelling is widespread. Starting with the availability of abundant computing power, college students often used high order polynomials to nicely fit their measurement data but with which you also could easily fit the coastline of Norway. As everyone

7 Nassim Nicholas Taleb: *The Black Swan*, ISBN 9780141034591, 2007

with a scientific background of course knows, most phenomena which originate from natural phenomena can be largely described by simple mathematical functions such as the square, the square root or the natural logarithm. So what does that nice fit actually mean? The compartmentalisation of science combined with the treacherous ease of numerical modelling has lead students to uncritically accept the results of the computations, especially when rendered in convincing powerful graphics. So it happened one and a half decades ago that one of my questions caused a student to redo part of his Master's thesis. The chemistry student had modelled the internal combustion process in a diesel engine and was interested in all reaction products entering the exhaust system of the vehicle. I asked him what the temperature scale was of the nice full-colour fringe plot of the modelled engine part. The highest temperature was many hundreds of degrees above the melting point of the engine material. Both the student and his supervisors hadn't checked with one of the mechanical engineers, amply present within the same laboratory. Nowadays, the issue of compartmentalisation has fortunately been solved to some meaningful extent, but the problem of a lack of healthy scepticism towards computed results proves to be persistent. A computer will always deliver numbers, either as plain numbers or dressed up as impressive plots. But the GIGO effect, Garbage In(put) Garbage Out(put), remains as valid as ever and remains one of its largest pitfalls. Many users are impressed with complex models requiring lots of data input, but you can question the usefulness of complex models if you need the data for say 30 parameters and you only know five of them with a high enough reliability. The other 25 have to be measured (if possible at all), estimated or even guessed. In line with the Pareto principle (see Chapter 2), more often than not it is much more practical to be roughly right (instead of precisely wrong) using a pragmatic five-parameter model describing 80% of the phenomena. If you lack sufficient reliable input required for the more sophisticated model, which in theory describes near 100% of the phenomena, you only get an illusion of accuracy. You risk fooling yourself (and others).

In Chapter 6, we will encounter an example of the misuse of models. The example there is the notion that reserves over production ratios for a range of metals stay more or less constant over decades (graph with horizontal tendency) but this doesn't say anything about the quality of the reserves in question or about the absolute quantities involved.

Another example is the misuse of inflation-corrected price graphs to explain anything you need: prices are such abstract data that most "re-engineered" explanations are completely unfounded. You can have an almost infinite supply of hypotheses leading to the same price graph. Another issue with the danger of almost unconditional faith in models is the validity of the underlying assumptions. Price models often assume a perfect market mechanism, whereas in reality markets are imperfect, as we saw in Chapter 2 with under-priced goods which will be over-consumed. Markets can remain irrational, e.g., an artificial price far below the real value of a good, for a long time until production can no longer keep up with demand due to over-consumption, after which the price will increase. This price increase could lead to a self-reinforcing feedback loop (see Chapter 4) by scaling up production to meet demand which again lowers the price and we start over again. Another phenomenon is the masking of scarcity by the focus on price issues: some rare metals are used in such tiny quantities in mass-produced products that the product price itself sets no limit on consumption, because no matter what the cost of the rare material itself is, it will hardly be reflected in the price of the product itself. But the sheer number of products produced and sold each year causes demand of the rare metal to exceed fresh supply. The mentioned phenomena are often not (or not well enough) accounted for in models.

I could go along with many other examples, including overstretched abstractions neglecting reality and including overstretched validity ranges (unauthorised extrapolations), but they all come down to an over-reliance on computer modelling. Too many people use or have to use models with underlying software that is so complex that they don't really understand how their problem is being computed. Besides small, fast, reliable and safe, software should be non-complex. And if a model is not good enough, it should be either refined or expanded without becoming incomprehensible or too complex, or it should be rejected. Computers may come in very handy, but are no substitute for independent thinking, (collective) experience and intuition and sound judgement.

ABUNDANCE THINKING

John Michael Greer[8] has coined and strikingly explained the notion of "the logic of abundance". The positive flipside of abundance thinking is its positive meaning with respect to the Elements of Hope, discussed in Chapter 7. Greer has also extensively discussed our predicament with regard to resource depletion in a number of other recent essays using the distinction between products and services derived from primary, secondary and tertiary goods.

Abundance thinking includes the notion that "the free market will solve scarcity" because our recent history of abundance has proven this notion to be true most of the time. Abundance thinking seems to be a firmly ingrained collective conviction. The biggest fallacy of abundance thinking is the often unquestioned belief of perpetual growth, or in a somewhat milder form the belief in aggregate perpetual growth interrupted now and then by a temporary dip. On a planetary scale it is easy to imagine and see that this cannot be true. So why do so many people still cling to this belief in growth? Maybe is has to do with our common notion of money. Primary goods are resources such as food, fresh water, fossil fuels and minerals from which we can extract metals and other materials. Secondary goods can be defined as products and services derived from primary goods by people for other people. Primary and secondary goods have physical limits. Enter the human abstraction of money and financial products.

THE MEANING OF MONEY

Unlike primary and secondary goods, tertiary goods, i.e., financial products have no physical limits. We can create money out of nothing (fiat currency), sometimes in the form of coins and billets but mostly in digital form. Money can grow very fast (say 10% annually) by issuing products like mortgages and government bonds. Banks typically lend out in the order of 10 times the amount of money they own or keep in store for savers (fractional reserve banking) and that's a good reason to dislike things like a run on the bank. And all the while lenders of money want to receive interest on their loans and this is why our economy has to grow all the time: besides the loan also the interest has

8 John Michael Greer, *The logic of abundance*, March 24, 2010, http://thearchdruidreport.blogspot.com/

to be paid back. Without growth, we have a serious problem with this system. For a number of generations now, many countries (including most Western countries) have experienced almost continuous growth, interrupted now and then by a temporary dip. But as they nowadays state as a disclaimer with financial products, "good results in the past are no guarantee for the future". The fundamental problem with our current system is the unchecked growth of the tertiary sector while at the same time the secondary system is failing to keep up due to lower-than-needed growth in extraction from the primary sector or due to a piecewise collapse of the primary sector caused by over-exploitation and excessive pollution. We are consistently mistaking money for the value it is supposed to represent. Money is a paper (and nowadays mostly a digital) promise for future products and services. What about the value of money if there aren't sufficient future products and services left to honour all these promises? And, in accordance with the observation that we tend to push the correct levers in the wrong direction (see Chapter 4), we are trying the remedy the financial problems at the turn of the first and second decade of this century by creating ever more money, illustrated by bailouts of entire countries. This doesn't bode well for our paper and digital promises, including saving accounts and pension savings.

So abundance thinking has seriously affected our vision on the meaning of money. The fallacy of thinking "if something threatens to become scarce, the market will always self-correct using the price mechanism" demonstrates this. This can be quite dangerous in obscuring resource depletion and scarcity by just looking at cost issues and by attaching too much meaning to them.

First, there are products using modern technologies enabling the use of very small amounts of relatively scarce materials. An example is gallium (^{31}Ga): a modern LED lamp replacing a classic light bulb needs only around a milligram of gallium, so the price of gallium can become ridiculously high before this will seriously impact the price of lighting hardware. Even with global shortages of gallium (demand exceeding supply) we could simply produce many billion LED lamps annually using gallium (around 1 ton gallium per billion LED lamps) because the lighting hardware will remain affordable. But applications demanding much larger amounts of gallium per product will then be priced out of the market (think large gallium-based photovoltaic solar panels). There are ample examples of other rare materials, often related to micro-electronics in both the luxury applications (nice to have) and in applications essential to our civilisation (need to have). So the irony is that the benefit of mass production combined with modern technology allows for cheap non-essential (non-survival) products for the global population, whilst at the same time making essential hardware prohibitively expensive for an increasing number of poorer countries. So the irony is that industry (in a holistic sense) makes more profits making photo and video making cell phones with touch screen, transparent mini-keyboard and wireless internet, using up many of the rare materials by the milligram, thereby making those same rare materials increasingly unavailable for essential hardware like medical diagnostic equipment and high efficiency photovoltaic solar energy harvesting equipment.

Second, a major part of the cost issue can be hidden from plain view because part of the true cost is not incorporated, e.g., fresh water consumption (often still "for free") or "free" dumping of waste in our environment. As we all know, low cost leads to over-consumption, be it unhealthy industrialised "food" leading to obesity in the US and other rich countries, or mass air travel fuelled by tax free kerosene and big efficient planes leading to price stunt airlines and rich northwest Europeans having bought a second home for the weekends in the southern or eastern parts of Europe (pre-crisis of course).

But the biggest fallacy of abundance thinking remains the fact that most of us mistake money for the value it is supposed to represent. And if we combine this with the notion that people are essentially herd animals, motivated by greed (food and in case of humans also money) and fear, we are on track of one of the root causes of evil.

Doesn't money mean anything then? Yes of course it does, but we should value the price relative to other goods and services as well as the rate of change of the price over the absolute price. Manufacturers or investors for instance may prefer to see a gradual and moderate increase of prices for raw materials instead of a high price volatility making planning ahead for these manufacturers or investors very difficult. Inflation an deflation should also be seen within the complete context. People living in communist European countries during the Cold War enjoyed an officially absent inflation during extended periods of time, meaning steady prices for consumer products, but that is of little use when the shelves in the shops are empty most of the time. And in times of deflation for instance, lower prices may be of little comfort or use when only few can afford to buy. So, again, prices do not always reflect if something is scarce or not.

THE REFUSAL TO ACCEPT PHYSICAL CONSTRAINTS

Another misleading result of abundance thinking is the common refusal to accept that physical goods can really get in short supply. Instead of analysing the data available and trying to understand the dynamics of the phenomena involved, the notion of anything that would contradict abundance is quickly dismissed. Scarcity of fossil fuels or certain minerals is often 'explained' solely in terms of so-called above ground factors (like geographic distribution and speculation) whereas in reality these factors do play a role but together with many other factors such as geology and physics. Peak supply is being denounced as being alarmist and replaced by the notion of 'peak demand'. This is the fallacy of trying to simplify real phenomena into an artificial distinction between cause and effect. Think of a loaded rope: it is rather useless to discuss the state of the rope in terms of the stress in the rope causing the strain in the rope or vice versa, the stress and strain emerge in each others company. There is no simple cause and effect relation with supply and demand either as tangible goods get scarce. Supply and demand are interconnected. Peak demand can be induced by supply constraints and supply constraints can be induced by peak demand. Supply and demand can decline simultaneously, but in the absence of sufficient affordable supply, a certain level of latent demand may loom in anticipation of more favourable times, resulting later in resumed resource consumption at maximum rate until the next decline.

GRAND GESTURE WITHOUT CONSTRAINT

Leaving out essential parts from the area of concern as a result of abundance thinking may cause pipe dreams like the sympathetic plan published by *Scientific American* in November 2009[9] to replace *all* fossil fuels by 2030. Most politicians love this kind of grand gesture stuff. The plan includes construction of 490,000 1MW tidal turbines, 5,350 100MW geothermal plants, 900 1,300MW hydroelectric plants, 3,800,000 5MW

9 Mark Z. Jacobson, Mark A. Delucchi, *A path to sustainable energy by 2030*, Scientific American, pp. 58-65, November 2009

wind turbines, 720,000 0.75MW wave converters, 1,700,000,000 0.003MW rooftop photovoltaic systems, 49,000 300MW concentrated solar power plants and 40,000 300MW photovoltaic power plants. Together. They don't explain where and how they are going to get the metals to construct all this, except put most of their hopes to recycling. They only mention a possible "materials hurdle" for a few already obviously constrained materials like neodymium, indium, tellurium and platinum. Otherwise, the plan seems to suffer from the conviction that materials are abundantly available at low cost, as was actually the case for many decades until recently. But these noble plans have to be literally materialised and as you will understand after Chapters 5 and 6, material supply is a major bottleneck, especially when it involves exotic materials required for high-tech high-efficiency 'sustainable' energy production. I will mention only one specific bottleneck here and that is the need for the rare earth metals neodymium (^{60}ND) and dysprosium (^{66}Dy) for 5MW wind turbines. These huge wind turbines need direct drives to skip the big and heavy gearbox that would be needed otherwise and which would give problems related to amongst others maintenance (especially offshore) and efficiency (friction losses and the need for cooling). The number of wind turbines to be built within the next 20 years requires some 3 million tons of neodymium whereas the current annual primary production rate is around 18,000 tons. Even if only part of the required wind turbines would use neodymium, there would still be a serious shortfall of this material. The required amount of dysprosium is much smaller, but dysprosium is much scarcer than neodymium and the current main producer, China, needs most of the production of these rare earth metals for its own industry and is considering a full export stop for dysprosium. For neodymium export restrictions are already in place for several years. Even if China would be willing to export all the neodymium (and dysprosium) it produces during the next 2 decades, either as a raw material or as finished goods for wind turbines, it is far from enough for the required number of wind turbines and it will not cover the demand from already existing large-scale applications such as hybrid cars or a plethora of other existing applications.

Similar supply constraints are applicable to all the other parts of the plan published by Scientific American. The biggest fallacy of this kind of grand gesture schemes is the provision of a false sense of security. It lures people into daydreaming about a hopeful and shiny future, a future for which we don't really need any sacrifice because there will be plenty for all and of course that is the kind of message that people want to hear and that politicians like to communicate. But we have to stop dreaming, wake up, face reality and let us guide by more complete and better assessments.

THE FALLACY OF CRADLE TO CRADLE WITHOUT CONSTRAINT

The long-known starting point of cradle to cradle (C2C), the notion that in nature each little bit of waste is eaten and used by something else, is of course well-founded. The cradle to cradle book[10] is important and inspiring but the C2C notion goes too far in claiming that we can consume as much as we want without feeling guilty as long as we do this 'cradle to cradle'. This is regrettable since the sound and valid base of the whole C2C thought will then ultimately lead to facilitating continued consumption growth, albeit in an 'environmentally friendly' way. Our world population still grows (currently by more than 70 million people per year) and the people who are already here want a higher standard of living. Without some sort of self-restriction, even with cradle to cradle, supply of primary and secondary goods will outstrip demand if we continue consumption growth. Think of energy, materials, water and food. A C2C approach is not the answer to everything and a C2C certificate is not a substitute for critical independent thinking. Sound and recommendable principles like C2C have been abused many times before, it happens now and it will happen again in future. I propose to pursue the sound basic principle of C2C within the broader context of the Elements of Hope combined with Managed Austerity (Chapter 7).

TECHNOLOGY WORSHIP

Technology and innovation will enable us to soften the consequences of global resource depletion (see Chapter 7), but don't get carried away with it. Don't expect miracles. Technology has given us tremendous benefits and with technology the magic box of tricks is far from being depleted, but it is by no means a cure for all trouble.

10 Braungart, M, McDonough, W., *Cradle to cradle*, 2009 (first published in the USA in 2002), ISBN 9780099535478

For starters, technology has to abide with the laws of physics. More specifically, you cannot beat the second law of thermodynamics using technology. The second law of thermodynamics means amongst others that energy conversions will always be accompanied by losses. It takes energy to concentrate energy. This is the reason why the claim, that the amount of solar energy hitting the Earth during a half hour is sufficient to power our global economy for a whole year, is true and at the same rather useless: the energy concentration difference with the ambient environment is low, very much lower than with fossil fuels. We need more materials per unit of installed power generation to harvest diluted forms of energy like sunshine and wind than for fossil fuels.

Timeliness and ability to scale are amongst the most important benchmarks to filter feasible propositions. Concepts that have been proven to function at a laboratory scale are often decades (not years) away from successful large-scale application. A lot of hurdles have to be taken (if possible at all) and that is why 'seed money' to fund innovation is so important: only a few percent of the many projects initiated (seeded) will prove to be able to mature into large-scale application.

Another important aspect to take into consideration is the fact that our technological solutions increase the risk and efforts related to the next level of problems. We are constantly raising the stakes. We use technology to produce oil from wells beneath miles of seawater, salt and rock and which is in some respects more complicated than space travel. And if things go wrong, we need an even higher level of sophistication to try to remedy the situation (if possible at all). A sad example is the disaster with the Deepwater Horizon drilling rig in the Gulf of Mexico in April 2010, resulting in a devastating gusher of oil and methane. Also, the stakes can become too high without technology being the constraint. Think of the dikes protecting parts of The Netherlands. These dikes often protect against a flood of water having a height difference with built areas of a few meters. Technology doesn't prevent us to build higher dikes to protect even against say 10 or 15 meters of water instead of a few meters, but nobody wants that. A dike breakthrough with housing a few meters below will result in substantial material damage, but most people have a good chance of survival and built areas may still be repairable afterwards. However, with a breakthrough at higher levels (say 10 to 15 meters) nearly everything will be lost.

We should also ask ourselves if we are making the right choices. More often than not, technology is not the problem. We already have the technology to build fully electric cars. But should we pursue that direction? As we will see in Chapters 5 and 6, the restrictions with regard to products like electric cars are the required energy production rates and the required rare metal production rates to realise large amounts of these products with the desired specifications at affordable prices. We need those resources for other products and services as well. If essential services like food and water, health, energy, communication, transport and law and order are at stake, maybe we have to reconsider if individual motorised mass transport at the current scale is even desirable during the next decades.

For some people, technology is like a religion. It comes with spokesmen who are almost treated as saints and their books are seen as holy works. It comes with conferences which sometimes feel like a church ceremony. You encounter this behaviour in various compartmental domains, including non-technical ones, seen by their worshippers as the one or at least as the main solution. If your line of business or research or consultancy is in wind farming, then wind farming is going to save the world. The same goes for solar energy, wave energy or some other specific technology. If you are a convinced, or even better, a converted (thus even more fanatic), proponent of biomimicry, then biomimicry is going to save the world. Of course biomimicry, like cradle to cradle, is based on sound and recommendable principles. Biomimicry is the notion that nature has optimised life during many millions of years and that we can learn from her (which is obviously true). Like C2C, it is a long-known notion, strikingly formulated in catchy language and more importantly accompanied by accessible explanations in valuable books and supported and expanded by new research. I want to remind you of the patterns discussed in Chapter 2. We are hardwired to recognise patterns, so please do not get carried away with a single pattern. If you see a certain type of beneficial design occurring in different domains within nature, you risk going too far in seeing application potential for this everywhere. But your gut instinct will undoubtedly say that this cannot be the magic solution for everything. Nature is a valuable and inspiring example. But don't worship nature without thinking: nature can be very tough. Eat or get eaten. We don't like earthquakes or hurricanes or other natural disasters. Nature doesn't negotiate. Our ancestors had short, hard and dangerous lives. Many people have a romantic perception of nature, but forget that they are looking at nature from the safe and comfortable viewpoint of our current industrial society. Nevertheless, we still are part of nature, we have largely forgotten this, and we desperately need to restore the balance.

The big pitfall of a strong concept working very well within a certain context is context dropping: over-extending the application of the well-functioning concept out of its valid boundaries. The best example is the conviction of technology worshippers that technology is going to save us. Didn't it take only about half a century to go from the flight of the Wright Brothers to supersonic jet flight? Didn't we succeed in getting men on the moon within a decade after setting this goal? Again, don't get me wrong here: we need more technology in the years ahead, not less. Just don't get carried away with it. We need technology which takes into account the scarcity of primary goods like harvested energy and concentrated metals. We cannot simply expect harvesting low exergy energy sources like sunshine and wind to make a similar leap as jet flight and space flight: they are bound by the laws of thermodynamics. Jet flight and space flight are of course bound by those same limitations, but we overcame those limits by using high exergy energy sources in the form of abundant fossil fuels. And we can no longer grow our consumption rate of fossil fuels (see Chapter 5).

I propose a simple cure from technology worship. Sometimes the easiest way to cure patients from a certain addiction is to let them change to a less harmful addiction. So maybe people who worship technology can worship thermodynamics instead, which will see to it that the many experts who propose plans to substitute fossil fuels with renewable energy schemes accompany their plans with the inconvenient truth that we need to simultaneously reduce our energy consumption. And while they're at it, the now thermodynamics-aware experts might as well admit that the many materials that are needed to establish a low exergy energy harvesting infrastructure are *not* abundant and that these materials need lots of exergy to be harvested themselves to start with.

Unfortunately, technology worship is linked with complexity worship. People and organisations mostly behave according to the way they are being rewarded. It's usually profitable to make things more complex. So our current industrial society has a strong inherent resistance to change into a direction of less complexity. Inertia is not just a law in physics. It is in the interest of people and organisations who benefit from our current high-tech high-complex society to carry on and further increase the level of complexity. But according to the laws of thermodynamics there are limits to complexity, dictated by the energy costs involved in growing and maintaining ever higher levels of complexity. This will be further discussed in Chapter 4.

MANIPULATED CONVICTIONS AND SELF-DECEPTION

One of the most perverse manipulated convictions is the notion of perceived obsolescence (besides planned obsolescence), ingrained in the minds of countless consumers by marketing and advertising as the notion that the products they own or use are falling out of fashion. This wouldn't have to be such a problem with proper second and third lives of products, but because we tax labour so high and 'value' resources so low, way too many products end up in waste dumps or incinerators prematurely. In the context of the ongoing global resource depletion, this waste of resources is a profitable form of stupidity. This logic causes manufacturers of mobile phones for instance to reduce the planned lifetime of cell phones for high school youth to a mere six months because it is being marketed as a fashion gadget so the supply chain can increase turnover rates and increase profit flows. And don't make the understandable mistake that responsible recycling compensates for this: the irony is that the energy required to make a new electronic product is generally much higher[11] (often many times higher) than the amount of energy saved by recycling the materials from which the product is being made. And recycling rates of many 'rare' materials are currently much too low anyway. So proper recycling of electronic products like cell phones may sound environmentally responsible (and it is), but it is much more effective to design and produce products like cell phones with a longer life. Of course this is easier said then done, because besides blaming marketing and advertising, we should blame ourselves as well because we show the corresponding behaviour and won't easily change that behaviour.

The debates on various important issues tend to be influenced by all kinds of interests, for obvious and partially good reasons of course.

11 Gutowski et al., *Thermodynamic analysis of resources used in manufacturing processes*, Environ Sci Technol., 2009

If you are interested in getting a clearer picture of the situation, it is recommendable to ask yourself a number of questions. Who is giving information and who supports this financially or otherwise? Who has to gain? What and how much is at stake? What about the timing? Is information being withheld or twisted or is partial false information accompanied by a partial truth to fabricate credibility? And is this ignorance, a mistake, pure coincidence or intent? Try to make a distinction between opinion and facts. Also be aware of other patterns of manipulating convictions. A well-known example is the way authorities can stall for time on a sensitive subject (e.g., climate change) by organising a conference with experts on the subject. Let's suppose that in the real world a large majority of the subject matter experts agree on climate change being real and a small part of them disagree. You can then invite (as actually happened years ago) roughly the same amount of speakers from both sides, have the proper (usable) chairman (either aware or unaware of the game) and then let him conclude that as long as the subject matter experts are in disagreement, authorities can not yet make a founded policy. It creates doubt in he minds of the general public and serves to establish controversies. And this is only the 'civilised' version of competing with facts and evidence. In politics as well as in science you can as a last (and actually desperate) resort try to discredit the bearer of the inconvenient message by lack of solid arguments. Perception management can be a very powerful weapon. Perception management cannot trump reality and will ultimately fail (and will often turn upon its users), but during the time gained in the process the damage is being done (and profits have been taken).
We should ask ourselves, on a regular basis, what our internal motives are. Let's be positive (or should I say naïve?) and suppose that all people, including all of us, have the best of intentions. We can then conclude, when we combine this positive angle with the text above, that we are deceiving ourselves on a large scale. And if we fool ourselves, others can fool us too easily. Now, that last sentence certainly doesn't sound so nice, does it? If we don't want to be fooled by others, we have to stop fooling ourselves. And if we want to stop fooling ourselves, we have to stop fooling others because most humans tend to believe their our own lies after a while. So besides noble intentions (all of us of course), also bare self-interest (everyone else of course) can act as a strong motivator to put an end to the manipulation of convictions in order to stop self-deception so that we won't be fooled by others.

One of the examples of manipulated convictions and self-deception is the fallacy of official policies with regard to resource depletion and the false promise of striving for fair distribution. Until now, official policies come down to the notion that resource scarcity is not so much a physical issue but is predominantly a matter of unbalanced distribution, ill-functioning markets, geo-politic considerations and so on. So-called above-ground factors are over-privileged in explanations and analyses at the expense of the physical constraints. And as one of the solutions to 'solve' part of the involved above-ground factors, policy makers and politicians are talking about balanced trade or fair distribution now that the global economic system, from which most of the industrialised countries have benefited from for many years, starts working against them due to depletion of the accessible resources within their own boundaries. Do we really want fair or equitable distribution? Because if we really mean this and we act accordingly, for the developed countries it would mean going back from first-world prosperity levels to third-world prosperity levels. Equitable distribution makes a noble concept in theory, and as far as practice goes it reminds me of the playful definition of a communist (except the communist party-leaders of course): someone who has nothing and likes to share this with others.
A more specific example is given by the conviction that we can continue driving cars at the current scale size and intensity by swapping internal combustion engine vehicles for fully electric vehicles. It gives the perfect example of overshoot in reducing complex problems to the extent that crucial elements are left out of the equation. People from industries and organisations with vested or future interests in building an new infrastructure with and around fully electric cars insist that there is no shortage of lithium to continue building and driving hundreds of millions of cars globally in the next decennia by changing from fossil fuel driven 'conventional' cars to electric cars using lithium based batteries. What these 'abundance thinking' and 'technology will solve everything' proponents obviously miss are several essential points.
First, they confuse quantity with quality. It is the never ending story of "there are billions of kilos or tons of such and so beneath our feet". This may be true and at the same time it's actually quite useless information, because it is the feasible production rate that matters, not total quantity, as will be discussed in Chapter 5. We have plenty of water in the Mediterranean or Atlantic Ocean and we have ample proven technologies to desalinate and pipe the water to the desert, so why isn't

the Sahara Desert green yet? The answer has already been given in Chapter 2, discussing the meaning and implications of the second law of thermodynamics.

Second, they treat a singled out problem as being totally independent from all other issues. "If we started an Apollo Program extracting and producing lithium from US (or non-Chilean or non-Bolivian) soils, we could have all the lithium we'll need" I recently heard one of the panellist say during one of the many conferences these days on "rare metals". I only had to wait a few presentations until another panellist made a similar statement on an "Apollo Program for rare earth metals to be extracted from US (or non-Chinese) soils". Obviously you cannot have several Apollo Programs being carried out simultaneously without neglecting other competing needs, especially in a period of limited, zero or negative real economic growth. It's a tempting game: let's solve any issue one at a time on paper, then add them up as if they were independent and, behold, all the world's problems have been solved!

Manipulated convictions and self-deception manifest themselves in the various one-size-fits-all 'solutions' to cope with the finite character of fossil fuel supplies. First we had the promise of nuclear energy, including nuclear fusion (still 4 to 5 decades away?) and a short excursion into the fantasy of cold fusion, all of them promising us nearly infinite energy supplies at low cost. Among the next saviours was the hydrogen economy; then the bio-fuels were going to save us, and the latest fashion is to make everything run on 'sustainable' electricity. What probability do you assign to large scale electrification as the 'final' solution? As we will see in Chapter 7, we need a multitude of adaptation directions and things like sustainable energy (solar, wind, and so on) are an essential part of that. There are no silver bullets. Silver bullets miss their moving target and fly by on the lost trajectory of trying to solve a complex problem with a single, easy, fast and wrong solution. But before we start discussing attempts to solve complex problems, we will first have a closer look at complexity in the next chapter.

CHAPTER 4
COMPLEXITY, STABILITY AND RESILIENCE

INTRODUCTION

General complex matters such as global resource depletion are too important to be explained by entrenched experts. You hear entrenched experts claim that big banks are "too big too fail", not only in the sense that they shouldn't be allowed to fail but also in the sense that according to their models it's extremely unlikely that they fail. And then they fail. The entrenched experts claim that "nations are too big to fail", until they de facto fail as well and have to be bailed out. The truth is that our global economy as a whole is already failing. There is not a single safe fiat-based currency in the world. There are simply way too few resources left to legitimise the amount of money running around in our fragile digital systems (almost all money is now digital). There are simply not enough tangible, real resources left to continue growth in the way that we used to have growth during the last many decades. Exponential growth gives a doubling time of resource consumption of around 70 years divided by the growth rate. In early 2009, the IMF once again stated that "a healthy world economy has a growth rate of 3% or higher". Three percent growth per year equals a doubling time of around 24 years. What do you think the odds are that the world will consume as much of a resource in the coming two and a half decades as in a large part of our history up to now? Another trillion barrels of oil? Another 100,000 tons or so of gold, another half billion tons of copper and another 40 billion tons of steel in the next 24 years? What about food and fresh water for a still relentlessly expanding global population? We have arrived at a point where our industrial global civilisation is so complex, that uncertainty (like stumbling on serendipitous inventions) is no longer helping us forward, but is actually slowing us down: the opportunities of throwing sand into the delicate machinery of today's complex arrangements are simply too numerous. It took just one Chernobyl incident to significantly derail the extension of nuclear power. It just took one Deepwater Horizon drilling rig incident to slow down production from much-needed non-conventional deepwater oil. Nevertheless, hope remains. It only takes a single spark of light to refute absolute darkness. I'll come back to hope, meaningful action and reality-inspired motivation in Chapter 7.

COMPLEX ADAPTIVE SYSTEMS, BOUNDARIES AND LEVERAGE POINTS

As we all know and understand, we have to make abstractions from the complex reality in order to be able to make sense of it all, but the danger of abstraction is choosing the wrong boundaries, i.e., leaving essential parts out of the area of interest. So we should regularly switch between the level of the details and the higher levels of connection and overview to check upon the validity of the system boundaries as defined earlier.
As already touched upon in Chapter 2 while discussing the Pareto principle, complex adaptive systems have leverage points: places in a system where a small change could lead to a large shift in behaviour. We do not always recognise leverage points: some of the most important decisions are those made only by default because we didn't recognise the lever being put into action. And we tend to push a recognised lever too often in the wrong direction as Donella Meadows[12] points out. A clear example is the whole list of growth-induced problems we are trying to solve with more growth. Another ironic example is farm-fishing as a 'sustainable' way of producing fish to 'save' the wild fish, while these farm fish consume around one third of all wild fish harvested worldwide. As explained by her, leverage points are often not intuitive since the combination of simple elements (stocks and flows) creates complicated and dynamically complex systems which are difficult to comprehend. A number of leverage points as described by Donella Meadows[12] are discussed below and have been provided with examples.

12 Donella Meadows: *Leverage Points - Places to Intervene in a System* (1999), accessed via http://www.thesolutionsjournal.com/node/419, October 2009

— **Buffers**

Among the key concepts she discusses are buffers and this concept is part of the adaptation framework (albeit the least sustainable part) as discussed in Chapter 7. As we already know from ordinary life, buffers (like fresh water towers) can stabilise a system by shaving off the peaks in demand and supply within a complex system and can stabilise a system by preventing (to a certain extent) supply disruptions.

— **Physical systems and information flows**

Another leverage point are the physical systems and their nodes of intersection. "Often, the physical rebuilding is the slowest and most expensive kind of change to make in a system"[12] and this is one of the major challenges we face when we want to make a transition towards a more sustainable industrial society. The energy, materials and efforts involved in replacing our fossil fuel based energy infrastructure alone are already mind boggling. We'd better use them with maximum efficiency while they last (and as long as fossil fuels are available). To do so requires global transparency of resources left (good quantitative data) and the sharing of the best technology available. More in general: it requires having complete access to the information flows. I would like to remind you of the example of the 7-crew spaceship (instead of the 7 billion crew Earth spaceship) of Chapter 1 to visualise the need for this. Would the rest of the crew accept it if one of the crew members were to withhold information or knowledge crucial for the survival of their ship?

— **Delays**

Another leverage point is the length of time of a delay relative to the rates of system changes. "It's usually easier to slow down the change rates than change the delays"[12]. So if we run for example into a sustained supply shortage of a certain metal, and we cannot do much about the delay because it takes 7 to 10 years to build (from scratch) new additional mining and production facilities, we'd better slow down demand for primary production of the metal in question in the meantime, for example by intensifying recycling or by looking for substitute materials (see Chapter 7).

– **Balancing feedback loops**

"A complex system usually has many balancing feedback loops, so it can self-correct under different conditions and impacts. Some of these loops may be inactive much of the time, like the emergency cooling system in a nuclear power plant, but their presence is critical to the long-term welfare of the system. One of our big mistakes is to strip these 'emergency' response mechanisms for the sake of efficiency and profit maximisation. In the short term, we see no effect from doing this but in the long term we drastically narrow the range of conditions over which the system can survive"[12]. Nature's way of assuring resilience of ecosystems is biodiversity and maybe we should take the example of a system, which has optimised its long-term functioning over many millions of years, much more seriously.

"The strength of a balancing feedback loop is important relative to the impact it is designed to correct. If the impact increases in strength, the feedbacks have to be strengthened too"[12]. Balancing feedback loops are part of the solution or adaptation framework as discussed in Chapter 7.

– **Reinforcing feedback loops**

"A balancing feedback loop is self-correcting, a reinforcing feedback loop is self-reinforcing"[12]. A good example of a reinforcing feedback loop is the amplification of big movements in share values at the stock exchange caused by herding behaviour of investors, a phenomenon we already knew from the tulip mania in The Netherlands in the seventeenth century. "Reinforcing feedback loops are sources of growth, explosion, erosion and collapse in systems. A system with an unchecked reinforcing loop will ultimately destroy itself. That's why there are so few of them. Usually a balancing feedback loop will kick in sooner or later. Reducing the gain around a reinforcing loop, i.e., slowing the growth, is usually a more powerful leverage point in systems than strengthening balancing loops. Population and economic growth rates in the World model of Limits to Growth are leverage points because slowing them gives the many balancing loops time to function"[12]. Of course population growth is still a political taboo. The dark irony is that the more people we are able to feed until we hit the physical limits associated with food and water, the more people are going to suffer. After reading the rest of this book, maybe you share my scepticism about the obligatory extrapolation of world population towards 9 billion people around 2050.

— **Rules**

Changing the rules is a powerful leverage point. An example is the large impact that industrial designers can have by designing products differently, maybe radically differently, e.g., using only the Elements of Hope (see Chapter 7) wherever possible. Products should be fit for extreme (overall 90%) recycling and for a long life based on high quality and intentional reparability instead of planned (or perceived) obsolescence.

— **Self-organisation**

Resilience has many definitions in many different domains and can be defined amongst others as the ability to return to an acceptable (or even normal) level of functioning after serious disruptions. So stability strongly benefits from resilience. "The ability to self-organise is the strongest form of system resilience"[12]. In analogy with the results of self-organisation within nature, we should strive for a biodiversity of ideas and adaptation directions instead of a monoculture of approaches. We will then have plenty of options and possibilities to draw upon in responding in a flexible way, regardless how the future evolves.
Please note that adaptation can be pro-active as well as reactive. Pro-active sounds better than reactive thanks to the decades-long overdose of the term pro-active in anything from job opening texts to function profiles. There is nothing wrong with 'reactive'. It is a matter of sound and healthy adaptability to react as soon as things stop going according to plan.

COMPLEXITY AND THE "LAW OF DIMINISHING RETURNS"

As aptly formulated by Ugo Bardi, "in a complex system there is no simple cause and effect relation. Everything affects everything else and you need good quantitative data to understand the weight of all factors involved"[13]. Unfortunately, good quantitative data is what we are missing most of the time. "Causes are difficult to pin down. Complex interdependencies exist between causes, causes and effects are blurred and complex feedback loops distort inputs. Certain characteristic

13 Ugo Bardi, *Peak soil*, http://europe.theoildrum.com/node/6189, 2010

patterns recur, but with infinite and unpredictable variety"[13]. Events cannot be accurately predicted (as discussed in Chapter 2) and we'll never fully understand very complex systems, so we should always keep a wide margin of safety.

So the challenge is to try to comprehend and encompass all major factors involved to avoid oversimplification, because oversimplification or missing one or more of the major factors involved often leads to sub-optimisations which can do more harm than good.

Complexity costs energy, so complexity emerges only in societies that have energy to spare (simplicity requires less energy). Investments in complexity yield diminishing returns. Our current level of complexity requires enormous amounts of energy to maintain. As energy becomes more scarce and expensive, society will simplify itself. This will be accompanied by a reduction in complexity. If this simplification goes fast we call it collapse. Evermore complexity, amongst others in the form of technology, will not solve our current problems. A lack of technology is not the real problem and pursuing ever more complex technology is a typical example of pushing a correctly recognised lever into the wrong direction.

Complexity makes our society vulnerable because when systems are highly complex, individuals matter[14]. "If during a pandemic key workers fail to turn up, we could have failures in vital sectors like energy, transport, food, water and telecommunications"[14]. Our current supply lines are long and thin and dependent on frequent small deliveries. Supermarkets and big box stores measure inventory turnover in hours, not days. We are highly dependent on a constant flow of vital resources. Complexity also leads to vulnerability because "as networks become ever tighter in connectivity and speed, they start to transmit shocks rather than absorb them"[15]. Redundancy is eliminated to maximise profits, but over-centralisation can create a single point of failure. Mass production maximises efficiency, but minimises resilience. Nature's way of assuring resilience of ecosystems is biodiversity, the contrast with our industrial agriculture based mono-crops could hardly be larger. Unspoiled natural habitats can generally cope much better with environmental shocks than the ones that are cultivated. Non-resilient,

14 Debora MacKenzie, *Will a pandemic bring down civilisation?*, New Scientist, 2008

15 Debora MacKenzie, *Why the demise of civilisation may be inevitable*, New Scientist, 2008

fragile systems are prone to rapid devolution or collapse and are only sustainable in the short-term and at great cost. Stable systems on the other hand are highly diverse with weak connectivity. Maybe we should try to match nature's very long experience to strive for optimisation rather than trying to 'beat' her with our striving for maximisation.

As aptly formulated by Richard Koch[16], there is a natural tendency for business and our civilisation as a whole to become over-complex. "All organisations, especially large and complex ones, are inherently inefficient and wasteful. Complexity and waste feed on each other"[16] and this goes largely unnoticed as long as we dispose of cheap abundant energy and as long as we can neglect the consequences of our pollution. Progress from a holistic point of view ('overall progress') requires simplicity; and "simplicity requires ruthlessness"[16]. This helps explain why "simple is as rare as it is beautiful. Complexity slows down simple systems. Big is only ugly and expensive because it is complex. Big can be beautiful. But it is simple that is always beautiful"[16].

I consider it quite likely that we will see the following management style trend taking shape within the next years: a military inspired no-nonsense approach, adapted to the pressing needs of our civil society. There are various factors which may lead to such a development: the need for triage-like decision making and subsequent action as events unfold, the increasing public desire for strong leadership (amplified by the current general lack thereof) and the unambiguous nature of the military command structure. The simple (non-complex) command structure is stable and resilient because trust is given to the lower ranks to make independent judgements and decisions if and when necessary. Control is an issue for after the action. Higher ranks can draw upon the knowledge and advice of their subordinates, but they make timely and informed decisions based on sound skills, training and experience, not necessarily by consensus. As the saying goes, if you have to steer through turbulence, you should sail faster than the current. Governance based on a more autocratic leadership style has a definite advantage in this respect. Of course the military inspired no-nonsense approach will only function if it is not being watered down by a host of heavily constraining 'rules of engagement' and by ambiguous command and control interference by other organisational

16 Richard Koch: *The 80/20 Principle*, ISBN 1857881680, 1998

branches competing for authority. The core command and control structure should be kept smart and simple (as well as accountable) and we should be frugal and selective with adding 'supportive' structures. The prospect of more autocratic leadership is both an opportunity and a threat. As most things in life, it can be put to good use and it can be abused.

THE SECOND LAW OF THERMODYNAMICS REVISITED

In Chapter 2, it was explained that the laws of thermodynamics mean that a system left by its own strives for a minimal free energy content which in turn is a basic driver of self-organisation. Self-organisation can be a feature of all complex adaptive systems. Each part adapts to its local environment along the path of the least effort. What emerges is a macro structure without intentional design. We witness the results in the generous beauty of nature in general. Examples range from the splendour of mineral crystals to the formation movements of flocks of birds or schools of fish. So you see that the strive for a minimal free energy content, which is the combination of the strive for maximum entropy and the strive for minimal energy content and the most stable state, can lead to things that are simply beautiful.

We see self-organisation also emerge in many aspects of our human society. But we humans are not satisfied with the natural pace of our ambient environment. That wouldn't be such a problem, if it wasn't for the fact that we are in fact *'rate addicts'*. Where nature tends to optimise, we humans generally tend to maximise. It's never enough. We want more, we want if fast, we want it now. And the higher the rate, the higher the difference in energy concentration between our activities and the ambient environment, leading to excessive exergy losses. Moreover, high rate activities and associated hardware and systems tend to be very complex, resulting in excessive waste and pollution. So our rate addiction not only hastens our encounter with the limits to available high exergy energy sources (fossil fuels) but also speeds up our collision with the consequences of abusing our environment as a giant waste dump, resulting in the destruction of our habitat, loss of biodiversity and destabilisation of our climate.

The laws of thermodynamics also imply that the more concentrated an energy source is, the less abundant it will be (and vice versa). That is why large amounts of useful concentrations of fossil fuels and metals are only found in specific forms (oil, gas, coal, rich metal mineral ores) in specific locations. But it isn't enough to have large absolute quantities ("the Earth's crust is so big") and to have all the technology you need in place. If you don't believe that, then ask yourself once again why the Sahara Desert isn't green yet, for we have plenty of water in the Mediterranean Sea and in the Atlantic Ocean and we have multiple proven technologies to desalinate seawater. The next Chapter (Chapter 5) will discuss and clarify that global resource depletion is often not so much the problem of a lack of total quantity, but the problem of a lack of sufficient quantity per unit of time (*rate*) in the desired quality.

CHAPTER 5
ASSESSMENT OF LIMITS ON RESOURCE PRODUCTION RATES

INTRODUCTION

Energy production and consumption per capita as a mean global value has been flat for three decades now and will soon drop year after year. On a global scale we are facing energy scarcity due to the fact that the growth in production of fossil fuels can no longer keep up with the growth in global population. We are past, at or near the point in time where net energy production, which is largely based on fossil fuels, can no longer grow and will soon decline. This will worsen energy scarcity at an accelerating pace.

As a direct consequence of energy scarcity, the primary production of metals will be restricted more and more by energy constraints, since the concentration of metals from mineral ores is highly energy-intensive. Also other factors play an important limiting role in the production rates of metals, amongst others the need for large quantities of water which can be translated into an energy constraint as well. So energy scarcity means metals scarcity, or more in general materials scarcity (most elements are metals). Materials scarcity in turn aggravates energy scarcity because all energy sources less concentrated then regular oil, regular natural gas and hard coal, which are all currently being depleted, require even more energy and materials to harvest, transport, process and convert.

ENERGY SCARCITY AS CENTRAL CONSTRAINT

Richard Duncan[17] has illustrated energy scarcity in a very clear way by making a time graph of energy production (and consumption) per capita, i.e., annual global energy production divided by the world's population. Figure 1, after Richard Duncan, shows a rapid growth in energy production per capita (in barrels of oil equivalent per human) from the end of World War II up to the end of the 1970s. Since 1979 we are on a sort of plateau: energy production per capita no longer grows. Annual global energy production (fossil fuels, nuclear, hydro, etc.) kept on growing but not enough anymore to offset the world's population growth. The graph of figure 1 up to including the complete plateau region until circa 2004 is history and, except for detailed discussions whether the line describing the plateau region should be a few millimetres higher or lower in certain calendar years, the course of the graph until recently is a given fact. Data on the last few years is not incorporated in figure 1 yet, so this part is marked with a question mark. It's possible that during the last few years we had a spike in energy production per capita after 2004 (higher than in 1979), followed by a decline soon thereafter due to the global economic recession which started around 2008.

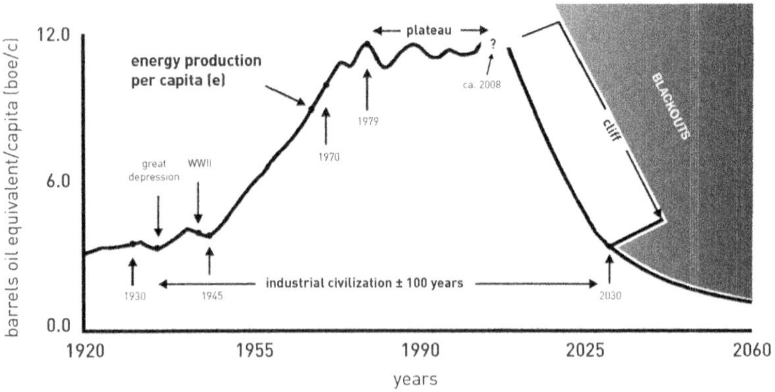

FIGURE 1

GLOBAL ENERGY PRODUCTION PER CAPITA AFTER RICHARD DUNCAN[17]

17 Richard C. Duncan, *The Olduvai Theory - Energy, Population, and Industrial Civilization*, The Social Contract, Winter 2005-2006

Of course, meaningful discussion should focus on the future course of the graph. The plateau region means that we had, on a global scale, a zero sum game with regard to energy per capita during the last 3 decades: when some part of the world's population enjoyed an increase in energy per capita in this period, this was at the expense of other parts of the global population. If the future part of the graph turns out to be correct, i.e. a (rapid) decline after the first decade of the twenty-first century, we will have a situation which is much worse than a zero sum game: even keeping energy production or consumption per capita flat for a certain part of the world's population will be at the expense of other parts of the global population. This doesn't bode well for the prospect of future conflicts. Of course it may well be possible to put positive feedback loops into action and innovate our way as much as possible out of the dark hole envisioned in the right part of the graph in figure 1. But given the fact that despite all our efforts and technology gains we haven't been able to grow our energy production per capita for three decades now, what do you think of the likelihood of a sudden increase in energy production per capita against the background of a relentlessly growing global population? We probably will be able to kick back the downward black line in the graph of figure 1 (towards 2030) to a trajectory of less rapid decline, but there is not a single well-founded reason to assume the plateau region to extend into the 2020s, let alone grow again as in the period before 1979.

Humanity has depleted a significant part of its inheritance of highly concentrated energy resources in the form of fossil fuels (oil, gas and coal). Although huge quantities of these resources remain untapped, the worldwide extraction rate (production flow) of oil and gas has reached a plateau and will soon begin to decline. The result will be a widening supply gap because sustained global economic growth requires sustained growth in available energy.

Oil and gas are currently the world's most important energy sources. Transportation for instance is currently almost entirely dependent on oil. Figure 2 gives the general depletion picture for oil and gas in absolute terms (in billion barrels of oil equivalent) and the left part of the graph strongly resembles a logistic curve. The initial stage of growth is approximately exponential, growth slows as saturation begins ("the low-hanging fruit has been picked") and at maturity growth stops and a maximum is reached. The maximum production rate is referred to as the 'peak' and is not a sharp deflection point in the curve but rather a plateau region. It is important to realise that the peak date in the depletion graph (figure 2) is not the same as the half date because production can continue for a long period after the peak. The actual depletion curve will presumably be asymmetric, having a peak date before the half date. Although the exact peak date for oil and gas is being contested (ranging from 2005 to somewhere during the next few decades), experts and authorities seem to converge on a peak date within the next years.

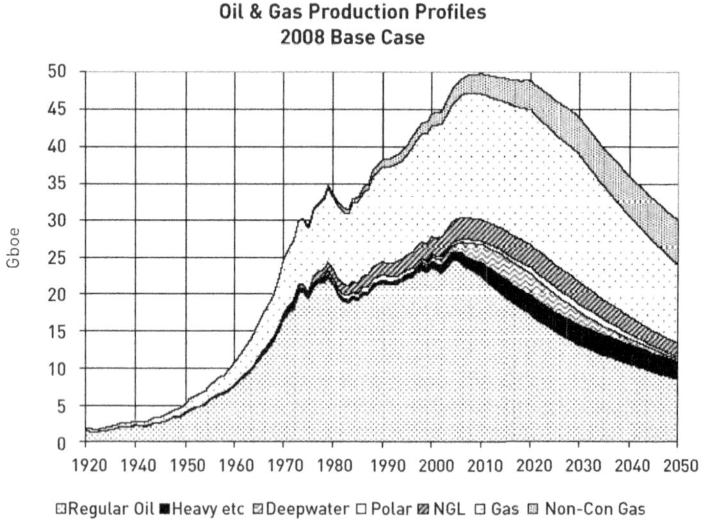

FIGURE 2

DEPLETION CURVE FOR OIL AND GAS ACCORDING TO ASPO[18]

Coal will not be able to fill the energy gap after the peak in oil and gas. According to the European Watch Group (EWG)[19] and according

18 Association for the Study of Peak Oil and gas (ASPO), *Newsletter No. 97*, compiled by C.J. Campbell, Staball Hill, Ballydehob, Co. Cork, Ireland, January 2009

19 EWG, *Coal: Resources and Future Production*, EWG-Series No 1/2007, Ottobrunn, Germany, March 28, 2007

to the "default scenario" in Richard Heinberg's recent book,[20] global coal production may peak around 2025. Again, this does not imply exhaustion of coal reserves, it is quite possible that more coal will be left for extraction after the peak date than has been extracted in total in the years before. The crucial point is that a maximum production rate will be reached after which supply can no longer grow and no longer follow demand. It is estimated that oil, gas and coal combined will reach their 'peak all fossil fuels' close to 2020[21].

Low quality fossil fuels like tar sands, besides having similar constraints as oil, gas and coal, suffer from a lack of energy concentration and have an accordingly low ERoEI (Energy Return on Energy Invested) as discussed amongst others by Charles Hall et al[22]. Water usage can also be a serious constraint, both as a required input and as a waste stream to be treated and carefully disposed of. In case of hydraulic fracturing used to produce shale gas, large-scale contamination of ground water has to be prevented. For 'regular' onshore oil produced in the USA, the ERoEI has diminished from around 100:1 in the 1930s through around 30:1 in the 1970s to around 12:1 currently. The ERoEI for deepwater oil (USA) is estimated to lie between 5:1 and 4:1 (before the Deepwater Horizon disaster). For Canadian tar sands, estimates for the ERoEI vary but are generally considered to be approximately 3:1, meaning that for every 3 barrels of oil equivalent we produce from tar sands, we have to spend one during the process. Earlier discussions in Canada whether or not to build a nuclear power plant to steam the tar from the sand gives you an impression of what such a low ERoEI actually means. Studies to quantify the ERoEI for shale gas are ongoing but will of course be lower than for 'regular' natural gas which is easier to produce. All other (non-fossil) energy resources combined (nuclear, hydro, wind, solar, bio-fuels, tidal, geothermal and so on) cannot fill the supply gap. The book from David MacKay[23] is one of the valuable

20 Richard Heinberg, *Blackout – Coal, Climate and the Last Energy Crisis*, ISBN 9781905570201, 2009
21 Sousa, L. de, Mearns, E., *Olduvai revisited 2008*, posted February 28, 2008 at the website The Oil Drum: Europe
22 Charles A.S. Hall, Stephen Balogh and David J.R. Murphy, *What is the Minimum EROI that a Sustainable Society Must Have?*, Energies 2009, 2, pp. 25-47
23 David J.C. MacKay, *Sustainable Energy – Without the hot air*, ISBN 9781906860011, 2009

sources available on this topic. Timely and massive utilisation of these other energy resources is limited by various constraints like lack of concentration, intermittency, issues related to conversion and storage and last but not least the required massive input of fossil fuels and materials to realise a replacing energy infrastructure. And materials scarcity is a crucial constraint as will be explained further on in this chapter.

For all these reasons we will probably be confronted with a peak in global energy production within the next 10 to 15 years, despite progress in technology. Other factors then the ones mentioned above, e.g., regional conflicts in production or transit regions for fossil fuels or production rate restrictions for geo-politic or geo-economic reasons, will likely expedite the peak in global energy production. This will result in a longer production tail in the future (more left for future extraction) but this is of little comfort as growth of our global economy needs growth in energy production. Local shortages will precede global shortages, meaning that timing and impact of 'peak energy production' will be different depending on where you live. Parts of the world will have their de facto 'peak energy' a number of years before we are at global peak. This is especially true for energy importing nations as energy producing nations may impose export restrictions to serve their own interests.

SCARCITY OF METALS

Metals have to be concentrated from mineral ores. Aluminium for instance has to be produced from the mineral bauxite which contains aluminium oxide (alumina). The depletion curve for this primary production (i.e., production from ores) will in essence follow a course like the curve in figure 3. The production quantity (vertical axis) is plotted against time (horizontal axis) so that we get the production rate of a certain metal over time. In reality a production profile will of course have ups and downs superimposed on the smooth curve shown here, but this is the common (simplified) picture. After a period of exponential growth (left part of the curve), maturity sets in because of the "law of the diminishing returns". When the low hanging fruit has been picked it takes more and more effort to concentrate metals from less accessible and lower quality ores (amongst others lower ore grades[24]). Due to this depletion of resources of higher quality and better accessibility, ever increasing amounts of energy and capital investment are required to try to sustain, let alone grow, production rates. At some point in time, production can no longer grow and a maximum, a plateau region, is reached. In case of a short lasting plateau we could call this maximum a 'peak'. After the plateau or the peak, production rates will fall. It is important to note that the area under the curve in part B of the graph (after the peak), representing total potential future production, is much larger than the area under the curve in part A of the graph, representing total past production. This will agree with your gut feeling that most of the relevant mineral ores will still be left under the ground after peak production has been reached. But as the saying goes, "it's not the size of the well that matters but the size of the tap". So the problem is not one of total quantities, but a matter of how much of those quantities can be converted into production per unit of time and at what cost.

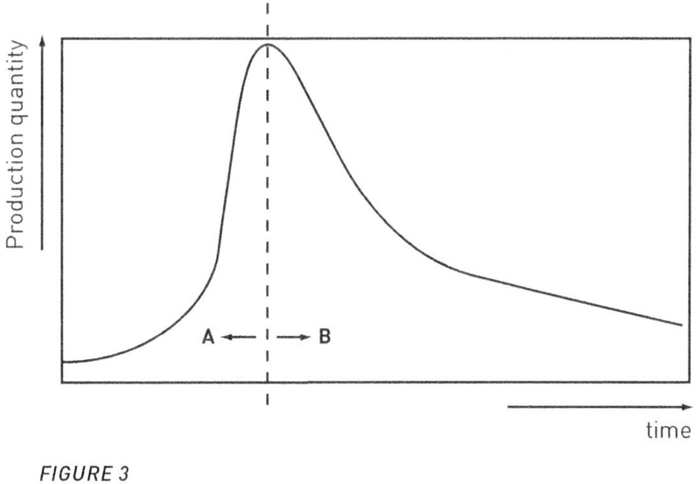

FIGURE 3

TYPICAL (STRONGLY SIMPLIFIED) DEPLETION CURVE FOR THE PRIMARY PRODUCTION OF RESOURCES LIKE METALS

24 See for instance figure 73 in *The Sustainability of Mining in Australia - Key Production Trends and Their Environmental Implications for the Future*, Gavin A. Mudd, October 2007

Many warnings in the past of impending metal minerals shortages have been proven wrong because of the availability of cheap and abundant fossil fuels. Every time the ratio of reserves to production of a certain metal containing mineral (or "metal mineral") became uncomfortably small, the reserves of that mineral were being revised upwards because it became economically feasible to extract metals from the so-called reserve base or resource base. Reserves are defined as those ores that can be economically extracted at the time of determination and the term reserves need not signify that extraction facilities are in place and operative. The decades-old paradigm which states that reserves will be increased (to include lower ore grades) as soon as supply gaps are looming, is no longer valid without cheap and abundant energy. Mining and extraction (concentration) consume huge amounts of energy; that is why mining is also described as "a way of converting energy into metals". The energy required for extraction grows exponentially with lower ore grades and this has been illustrated for instance in figure 3 in the paper from Page and Creasy.[25] This is also being reflected in figure 6.8 in the recent book from Ashby[26], giving embodied energies of produced metals which are typically several hundreds of Mega-joules per kilogram. Please note that the figures given by Ashby are not static. Ashby's book gives lower embodied energies than similar books a couple of decades ago thanks to technological progress with associated efficiency improvements. Future numbers will likely give higher numbers again as a consequence of less accessibility of ores (longer distances, less favourable climate conditions, more overburden, deeper mining), lower ore grades (exponential increase in energy input) and less favourable ore compositions (more difficult processing).
The highest ore grades have already been depleted or are already being mined. The United States of America, in fact half a continent in size, have had their peak in primary production of most metals before the 1980s[27]. These include not only silver (1916) but also widely applied metals like chromium (1956) and zinc (1969). This does not mean of course that you

25 Norman J. Page, S.C. Creasy, *Ore grade, metal production and energy*, Jour. Research U.S. Geol. Survey, Vol. 3 No. 1, Jan.-Feb. 1975, pp. 9-13
26 Michael F. Ashby, *Materials and the Environment - Eco-Informed Material Choice*, ISBN 9781856176088, 2009
27 Chris Clugston, *Continuously Less and Less - The New American Reality*, October 2009

couldn't mine and produce chromium and zinc anymore in the USA, but it means that it's partly not worthwhile anymore to do this in the USA. A similar pattern of peak production, in another order and timing than for the USA, will manifest itself on a global scale.

The absolute amounts of various metal minerals in the earth's crust are large beyond imagination, but the bulk of it might as well not be there because it's out of reach because of the required energy involved in extraction. Most of the minerals occur in practically useless low concentrations: below the so-called mineralogical barrier[28] (a certain low ore grade), essentially you should pull the source material (e.g., a piece of rock) chemically apart to extract the individual metals. This is of course prohibitively energy intensive. This energy restriction is strongly amplified (leveraged) by our predicament with regard to current and future energy scarcity. It is therefore very doubtful that meaningful parts of the reserve base or resource base of many metal containing minerals will ever be upgraded to reserves. It is even questionable whether all currently stated reserves are fully exploitable given the fact that the era of virtually unlimited amounts of cheap energy lies behind us. It's quite likely that overstatements of metal minerals reserves will be revealed in the not too distant future, as has happened in the past with oil reserves. Reserves have been used (and are still being used) as accounting terms and have been revised upwards (and are still being revised upwards) based upon anticipated demand.

Metals scarcity will be aggravated by other factors. The industrial revolution started in Europe and later the USA became an industrial giant, so it comes as no big surprise that both Europe and the USA have depleted a large part of their accessible resources. More often than not, the largest part of global primary production takes place in just a few countries (in accordance with the Pareto principle, see Chapter 2), often outside the Western industrialised world. Metals are obtaining a stronger geo-political and geo-economic meaning, just as fossils fuels, and can and will be used as a means of power politics. Most of the 'regular' oil left can be found in the Middle East (Saudi-Arabia, Iran, Iraq, Kuwait and the United Arab Emirates) and most of the 'regular' natural gas left can be found in Russia, Iran and Qatar. Currently, China

28 Brian J. Skinner, *Exploring the resource base*, workshop "The Long-Run Availability of Minerals", Washington DC, USA, April 22-23, 2001

is the world's leading producer of 16 metals[29]. Amongst them are the rare earth metals (97% of global primary production), tungsten (81%), barium (barite, 55%), indium (50%) and tin (45%). Other examples of the non-linear geographic distribution of global primary production of metals are Chile with 40% of lithium production and 34% of copper production, and South Africa with 79% of platinum production and 42% of chromium production. As with oil and gas, global ('average') shortages of metals will most likely be preceded by spot shortages because of geo-politic and geo-economic considerations, including export restrictions or even export stops. The analogy with a river is useful here, especially if river water supply (the flow rate) can barely (or can no longer) meet demand. If your country is located upstream, you have a beneficial position since you could decide to construct a dam to operate a hydropower plant and to have a water buffer to draw upon during dry spells or times of increased demand. If your country is dependent on the river but is located further downstream and more and more water is being tapped upstream, you are basically screwed without solid agreements with upstream neighbours, without money to purchase a certain water flow, without possessions upstream or without alternatives.

The peak in primary production of most metals will probably be reached no later than halfway the 2020s, mainly due to the global peak in net energy production. The peak in the primary production of most metals may be accelerated by an earlier energy peak caused by regional (or even global) conflict, by negative feedback loops during the onset of reaching peak energy production, by interferences with water shortages (see further on in this chapter) and by disruptive effects of large scale natural or manmade disasters. The peak may also be accelerated by a dragging financial / economic crisis, which is likely connected in a complicated way with energy scarcity as indirectly explained by the part "the meaning of money" in Chapter 3. An 'early' peak will lead to a longer production tail in the future (more left for future extraction), but this is of little comfort as we are interested in present production rates at affordable cost which we would like to see growing in order to grow our global economy.

29 Jon Price, *The Bright Future of Global Mineral Resources*, Nevada Bureau of Mines and Geology, 2010

ENERGY SCARCITY AND METALS SCARCITY REINFORCE EACH OTHER

Because of the depletion of fossil fuels, we need to produce energy from more diluted energy sources like sunshine and wind. The catch here is that the preferred and required metals to build the infrastructure needed to harvest and apply the alternative energy sources are quite restricted with regard to their production rates. These metals include the platinum group metals, most of the rare earth metals and various minor metals like gallium, germanium, indium and tellurium which are mostly dependent on other (base) metals for primary production. We are not able to increase primary production of these metals fast enough to satisfy surging demand. Their applications include high efficiency photovoltaic solar cells, permanent-magnet drives and generators (large wind mills, hybrid cars, electric cars), catalysts for cars and for petrochemical cracking, fuel cells, batteries and various electronic devices (from touch screens and hard-drives to energy saving lighting). Chapter 7 will discuss adaptation directions to cope with energy and materials scarcity. We will not be able to completely substitute energy production from fossil fuels by alternative forms of energy (not even close). But we can take meaningful action to increase alternative energy production as much as possible by focussing on less scarce materials, which have less stringent restrictions to feasible production rates, in order to literally materialise the required infrastructure.

But please don't forget the materials demand originating from our current energy supply and energy consumption infrastructure. The complete energy sector is the largest metals consumer, so metals scarcity aggravates energy scarcity whilst energy scarcity aggravates metals scarcity. The energy sector uses lots of metals within its whole chain of exploration, production, storage and distribution up to conversion into the desired forms of energy. Currently almost all energy is produced by either fossil fuels (most of it) or the 'conventional' alternative energy production in the form of hydropower plus nuclear power. All these methods of energy production require huge amounts of metals, the more common ones as well as the more scarce ones. The main constituent of steel (iron) is an abundant element, but many of the essential alloying elements are not. We need more and more infrastructure like huge offshore platforms and huge lengths of pipes. We need special steels and alloys for refineries and power plants, even more special in case of nuclear power plants. We need huge amounts of copper, both in producing and transporting energy and in using it. Whether we fuel our mobility (land, sea, air) with fossil fuels or with electricity, we can't do without the metals for the engines. All these applications have the competition of demand from other strong materials consuming sectors like building and construction and electronics.

Metals scarcity (like energy scarcity) poses a direct threat to our prosperity. Resource poor countries like most EU countries and like Japan have to import most of the metals input for their domestic industry.
Scarcity will manifest itself in one way or the other in cost, as already could be witnessed during the 2003-2007 timeframe. Price levels of many metals showed high spikes during that period. An extreme example is given by vanadium which price temporarily increased by around a factor of 20. But the absolute price level in itself is an insufficient indicator of scarcity. First the absence of price stability is probably much more important than the absolute price level: uncertainty is not very helpful in making investment decisions. Secondly, scarcity can also manifest itself at low price levels, meaning during a period of deflation. In the third place, money is rather useless of it can't buy you the necessary items: for a number of the rare earth metals, China is already using export quotas (since 2004) and considering export stops in the near future. China also put restrictions on higher volume metals such as magnesium, manganese and zinc, leading to a complaint of the USA and the EU at the World Trade Organisation in 2009.
History shows us that scarcity of various kinds often leads to conflict. It's no longer sufficient to look at the global geo-political situation from the perspective of the availability and distribution of fossil fuels, water and food; we should add metals to this list. Future access to raw materials will primarily be realised by 3 options: (1) by owning the raw materials or (part of) the production chains, (2) by using surplus of other products and services to be able to buy raw materials and associated products (or simply barter for them) and (3) by taking them by force. Option (1) could imply repatriating (part of) the manufacturing industries back to developed countries. Rich but resource poor countries have to utilise even more than now their inventiveness and creativity. It is a misconception to think that it is not much of a problem to deal with significant price increases in metals, because the value of the metals used in most cases presently represents only a fraction of the costs of

the finished products. We know from the Pareto principle (see Chapter 2) that a 49/51 division can quickly evolve into an 1/99 division (winner takes all) and this may also happen with competition from cheap but resource-rich countries, especially where crucial high-tech products and processes are involved. Just take a look at what happened and is still happening in China, which has recently surpassed Germany as the world's biggest exporter. A powerful way of avoiding the unbalanced competition involving scarce metals you don't have or you don't own is to try to use more abundant materials instead (see Chapter 7).

If possible, rich but resource poor countries will also have to put more emphasis on agriculture (food as well as non-food products) and, if applicable, a more intense use of their bordering sea or ocean. In this way, they can be more self-sufficient and create a greater surplus of domestic goods and services, option (2), to trade for non-domestic goods like metals and associated products.

The most important consequence of energy and metals scarcity is that this combination acts as a reinforcing feedback loop (see Chapter 4). Without sufficiently strong balancing feedback loops (proposed and discussed in Chapter 7), this will result in accelerating energy shortages as pictured in figure 1. The severity of the consequences can hardly be overstated. Until now, the availability of abundant cheap energy has acted as a very powerful leverage. A single barrel of oil contains the energy equivalent of more than 20,000 hours of manual labour. As a global average, every single human currently consumes around 12 barrels of oil equivalent annually (much more in the Western world). Take that energy away and our standard of living will drop beyond recognition; that is, if you live to witness this because world food production cannot be sustained at current levels without massive energy input (see next section).

SCARCITY OF WATER AND FOOD

Global resource depletion, comprising amongst others the interlinked scarcities of water, food, energy and materials, is fuelled by many factors. The two most important ones are the ongoing global population growth and the striving for better standards of living for the billions of people already present on our globe. The most pressing problem in the

longer run will probably be water and food scarcity: you could make a living with (much) less energy and (much) less metals, but we can't survive without water and food.
Adam Nieman (see the internet) has made a striking photo composition (during or shortly before 2004) of our globe together with a little bubble on its surface which represents all the water of the world on the same scale as our planet. Pictures say more than a thousand words and this picture makes it clear that water is a finite resource on our planet. Of the little water bubble in this picture, 2,5% is fresh water, from which 2/3rds are frozen (mainly in the Arctic area and on Antarctica). The largest part of the remaining less-than-one percent is groundwater, partly to be retrieved by pumping it out of large underground lakes called aquifers. In fact, we are partly mining 'fossil' water by depleting aquifers originating from the last ice age. The rest of the fresh water is circulating in the form of clouds, precipitation and surface waters such as rivers.

The problem is that we use water at a faster rate than nature replenishes it for us. Agriculture and power generation are the world's largest users of water. Irrigated agriculture accounts for around 70% of the water used daily by people, irrigation systems tap rivers to the extent that they no longer reach the oceans. Industrial use, including power plants, oil refineries and ore refining industries, accounts for the majority of the other 30%. We are depleting aquifers worldwide to satisfy the enormous and growing demand for water. Where aquifers have been depleted, desalination plants are constructed on coastal areas to satisfy demand. In 2010, around 150 countries worldwide are operating desalination plants. But desalination of seawater consumes enormous amounts of energy and they require lots of metals and other materials for construction. In the Persian Gulf region, most desalination plants run on natural gas, recently causing the first gas exporting nation in this region to become a net importer for this reason. The metals and materials demand for desalination stems not only from the construction of the desalination plant itself, but also from the pumps and pipes required to transport huge quantities of fresh water inland (and pumping costs electrical energy). Transport by truck requires additional fuel. Besides agriculture and power generation, also the mining industry up to including the concentration and production of metals from the mineral ores is a huge user of fresh water. Metals typically require dozens of litres to around a thousand litres of fresh water for every kilogram of refined metal (see the corresponding graph in Ashby's recent book[30]). The big mining corporations are investing billions of dollars in desalination plants worldwide, amongst others along the coast of northern Chile and along the coast of southern Australia.

The direct and indirect impact of water scarcity on energy scarcity can be illustrated by the impact of water scarcity on hydropower. Shortages of hydropower have to be compensated by generating electricity in existing thermal power plants running on fossil fuels, resulting in increased demand for fossil fuels. And if energy supply cannot meet energy demand, large consumers of hydropower such as metals refiners will be shut off from power to avoid having to cut off the main residential areas from electrical power (rolling blackouts). Aluminium smelters in both Venezuela and China have recently (2009) been forced to (temporarily) drastically cut production for this reason.

Climate change also has a direct link to energy scarcity. A warming globe possibly tends to make dry areas drier (and wet areas wetter, see Chapter 6), leading to shortages of hydropower and lack of cooling potential for power plants (especially nuclear), often during the same hot periods when demand for electricity is surging because of the demand for air conditioning. This already causes local and temporary blackouts, in developing countries as well as in industrialised nations. More importantly, higher evaporation rates in hotter areas increase the need for irrigation, in turn increasing energy demand for water production (and agriculture accounts for around 70% of global fresh water consumption). And major river systems, especially in Asia, risk loosing their summer feed from glaciers as they continue to shrink year after year. The other side of the medal, too much water per unit of time by excessive rainfall, can have a direct impact on mining as well (besides damaging agriculture). This is illustrated by the temporary shutdown of some of the world's largest mines of copper (Chile) and coal (Australia) in recent years due to flooding.

Food production is directly impacted by water scarcity. It takes for instance around one cubic metre of water to produce 1 kilogram of wheat, either to directly feed a human or to feed cattle to produce beef. It also takes around one cubic metre of water to produce 1 litre of milk.

30 Michael F. Ashby, *Materials and the Environment – Eco-Informed Material Choice*, ISBN 9781856176088, 2009

It takes around 15 cubic metres of water to produce just one kilogram of beef.

Food production is directly impacted by energy scarcity as well. It typically takes 6 to 10 calories of fossil fuel derived energy to produce and deliver the equivalence of 1 calorie of food in the Western world. This is being caused by the energy demand for amongst others the production, distribution and application of fertilisers, herbicides and pesticides, the energy required to manufacture, maintain and operate mechanised farming equipment (like tractors) and the processing, packaging, distribution and cooling of food products. The 'Green Revolution' between World War II and the late 1970s, which has enabled our current population level, has been made possible not only by progress in technology but was also literally fuelled by cheap and abundant fossil fuel energy.

So, energy scarcity and water scarcity, which are interlinked, mean food scarcity.

There is also a link between metals scarcity and food scarcity by the competition for land use. There are various locations across the globe where authorities can choose between continuation of current land use, e.g., farming, and a more profitable use by mining the soil. Let's take the extraction of titanium containing mineral sands as an example, as this can require the stripping of all vegetation and subsequent removal of the lower layers of the soil. This isn't a problem in desert-like areas in Southern Australia, but it is in other parts of the world, e.g., India. Mining could bring in more money then agriculture but after a couple of decades the mineral sand has been removed and the prospect of farming is gone, probably for good.

Because of energy scarcity, the price of food can no longer be viewed as being independent from its energy content. Staples like corn can be used to feed humans and animals but can also be converted into bio-fuels to feed the enormous global population of internal combustion engines. Because of energy scarcity, also the price of bio-materials, for instance plastics from sugar beet and corn, can no longer be viewed as being independent from their embedded energy content. And bio-fuels and bio-materials of course need huge quantities of fresh water as well. So we have a host of competing needs from various sectors for fresh water which are partly contradicting in a way resembling a reinforcing feedback loop. We are suffering from energy scarcity, but if we try to remedy the situation by heavily investing in bio-fuels (and bio-materials) which require huge amounts of water and which suffer from a low Energy Return on Energy Invested (ERoEI, see section on energy scarcity), we dig our hole deeper and aggravate amongst others food scarcity. Moreover, for the production of bio-fuels and bio-materials to be sustainable, we need to replenish the soil with all the nutrients removed from it by harvesting the crops; otherwise, we are just 'mining the soil'. Of course the same applies to food production.

Other factors can and probably will worsen the situation with regard to food scarcity. The loss of biodiversity and factors like water stress and global warming weaken the resilience of ecosystems and make vegetation more susceptible to plagues like bark beetles infesting forests in the USA and Canada and wheat fungus threatening wheat harvests in parts of Africa and Asia. The sharp decline in honey bee colonies worldwide, whatever the causes, increase the risk of less 'industrial scale' harvests of food items like fruits, nuts, coffee and chocolate, thus risking to narrow down our future variety of affordable food.

Our planet has a finite amount of arable land and the available surface goes down in quality as well as in quantity. Food scarcity, linked to energy and water scarcity and even metals scarcity as explained previously, will be further aggravated by additional factors. Soil erosion is an important factor. Our industrial-scale agriculture removes topsoil at a much faster rate than nature plus fertilisers replenish it. This process is accelerated by excessive rainfall on rationalised patches of farmland (no or few obstacles like bushes or tree lines), washing away part of the topsoil. Other factors include urbanisation, salinisation (the accumulation of soluble salts in soil to the extent that soil fertility is being compromised) and desertification.

On top of this all, a growing number of experts warn of looming shortages of cheap and abundant phosphate fertilisers, essential for plant growth, based on mined phosphate rock. But these experts estimate the timing of reaching a maximum production rate from phosphate rock somewhere between around 2035 (or sooner by energy scarcity) and around 2070 (or even later). According to Liebig's Law of the Minimum (see Chapter 2), we should be worried right now about water scarcity as this is already the predominant limiting factor for plant growth (on a global scale). There simply is no substitute for water. As military planners say: "key to long term survival is short term survival" so we should focus first and foremost on water availability (and keep a keen eye on affordable availability of fertilisers).

A large part of our global population also depends on the oceans' fish for their nourishment. But stocks of fish in the world's oceans are being depleted by over-fishing ('mining the ocean') and the capacity of the ocean's ecosystem to self-heal in the very unlikely case that we immediately cease industrial fishing completely to let nature regain her balance within a period shorter than a few human generations is getting smaller by the year. Increased atmospheric CO_2-levels and global warming have already affected the average temperature and average acidity of the upper layer of the oceans. An increasing area around the coastlines suffers from lower oxygen levels, sometimes resulting in dead oxygen-free zones. Although the change in average temperature and acidity is not large yet, an additional danger lies in the likely non-linear nature of this change, resulting in larger change rates in future years.

This whole picture of increasing risk of food scarcity is obscured by certain perversities of our current economic system. An example is the loss of some 2,000 tons of cherries in Belgium during the summer of 2009 (by not picking them from the trees) because the farmers would get a mere 30 Eurocents per kilogram, insufficient to cover the production costs including harvesting the cherries. In the meantime, the average Belgian supermarket charged around 9 Euros per kilogram cherries, and in neighbouring Luxembourg the stores in the capital charged around 18 Euros per kilogram. But this type of example does not refute the notion of possible food scarcity on a global scale. The whole notion of global resource depletion is obviously not an exact science where it just takes one exception to the rule to disprove that rule. Speculation (or price manipulation) is of course part of the whole equation, but food scarcity can not be explained away by oversimplification and by declaring a factor like speculation to be the primary cause.

In large parts of the world, water is still artificially cheap. This holds the prospect of larger efficiency in water usage once the user has to pay the real cost of the water, on the other hand it holds the prospect of higher food prices and will put additional pressure on water-intensive crops. Food scarcity will manifest itself in surging cost of food (larger percentage of disposable income), amongst others resulting in the loss of social stability as already happened during the second part of the first decade of this century.

CHAPTER 6
OUR PREDICAMENT: OVERSHOOT AND COLLAPSE

INTRODUCTION

We have wasted valuable time and lost valuable opportunities for meaningful actions during the last decades. People like the authors of *Limits to Growth* apparently were too far ahead of the crowd. Do we wait until we have 100% proof of hazards with no fall-back strategy at that final moment? Will we start constructing a fire engine for a yet-to-be-established fire brigade only until some major fire breaks out in town? Of course not. But we refuse to apply the same logic to other credible threats. Military wisdom is partly applicable here for our society as a whole: "better safe than sorry" or "hope for the best, prepare for the rest". It is important to understand that we are already in overshoot (like Wile E. Coyote above a canyon in the Road Runner cartoons of the 1950s and 1960s), it's too late for crisis prevention. It's never too late though for proper actions. That doesn't mean that our current predicament is a 'problem' with a 'solution'; it's more a collection of dilemmas. Nevertheless, it's much better to strive for improvements, if only to mitigate part of the effects.

PRODUCTION RATES AND RESERVES OF METALS

Annex B gives an overview of current primary production rates of most metals, based on the data from the USGS (United States Geological Survey). All data are estimates since USGS does not dispose of all data from all metals from all countries worldwide. But to the author's knowledge the USGS data is the best available consistent dataset covering the whole globe and open for public. Annex B gives all production and reserves numbers in tons of metal content. This means for instance that the numbers for iron ore are translated into numbers for iron based on the average ore grade (around 50%), so 1 ton of iron ore is assumed to contain half a ton of elemental iron. Again, this indicates that Annex B gives *estimates*, with the purpose of getting a feel for proportions as well as absolute numbers (orders of magnitude).

The production refers to the primary production, meaning the 'new' production from mineral ores. It excludes production from recycling after consumer use. Annex B also gives the USGS numbers for reserves, defined as those ores that can be economically extracted at the time of determination and the term reserves need not signify that extraction facilities are in place and operative. From these numbers, the R/P-ratio, the reserves to production ratio, can be calculated which is a sort of 'abundance' indicator. But R/P-ratio's can be deceiving and should be interpreted correctly. First, our current economic system needs continued growth, reflected in a continued global average production and consumption growth of metals (so far). So instead of calculating a 'static' R/P, it is more realistic to calculate an R/P-ratio which accounts for a typical (modest) production growth of 2% annually (Annex B) to assess the feasibility of 'business as usual'. Second, the course of the R/P-ratio over time can be incorrectly interpreted as 'evidence' for abundance by neglecting the quality of the reserves. The quality issue, not reflected in the R/P-ratio, can be explained as follows. Suppose the R/P-ratio for the primary production of a certain metal is around 30 for several decades. Then what do you prefer? A million tons of reserves near the surface, in a high ore grade (high concentration), in a benign environment and close to available labour and infrastructure? Or would you prefer to have a million tons of reserves deep beneath the ground, scattered in low ore grades, at a remote location in a harsh environment? By neglecting the quality issue and by focussing on quantities alone, a false sense of abundance and security can be fostered. A good illustration is the 'news' in 2010 that Afghanistan is host of "a trillion dollars worth of metal minerals". The same 'logic' dictates that the Greenland ice sheet contains trillions of dollars worth of fresh water. The real issue of course is which production rates are feasible at what cost. Ask this simple question to people advocating ocean floor mining or even asteroid mining or 'mining the moon' and your simple question becomes an inconvenient one.

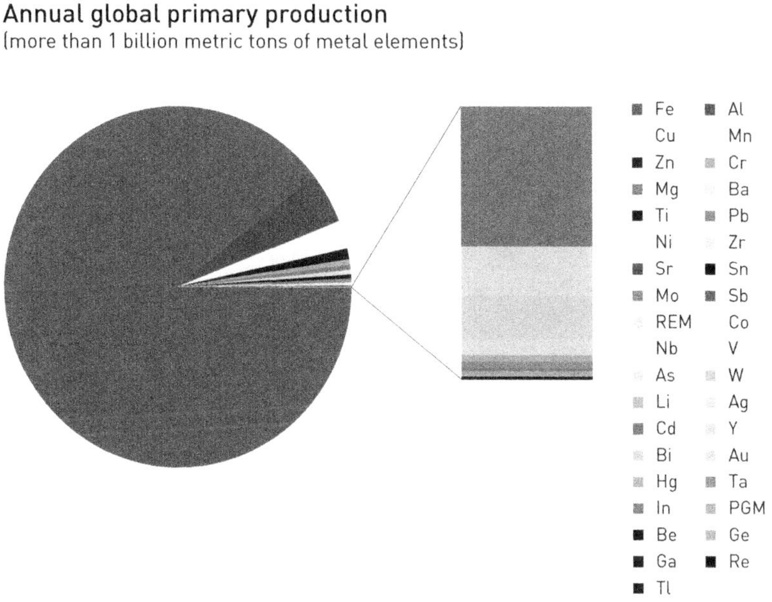

FIGURE 4

BREAKDOWN OF ANNUAL GLOBAL PRIMARY PRODUCTION OF METALS, BASED ON ANNEX B

Figures 4 through 7 depict in more detail global annual production rates and reserves, based on Annex B. The right parts of figures 4 through 7 give the metal elements in descending order of quantities; in figure 4 for instance, the largest part of the pie-chart is iron (Fe), followed by aluminium (Al), copper (Cu), manganese (Mn), zinc (Zn) and so on. The annual primary production of iron dwarfs all other metal elements combined. Despite its huge reserves, high-grade iron ore will last less than 50 years (at 2% annual production growth) as far as cheap and abundant primary production is concerned, due to the enormous scale of its annual global consumption. Magnesium reserves are virtually unlimited because of its abundance and associated accessibility in seawater (more than a kilogram per cubic metre of seawater), albeit at high energy cost for concentration from seawater compared with primary magnesium production on land.

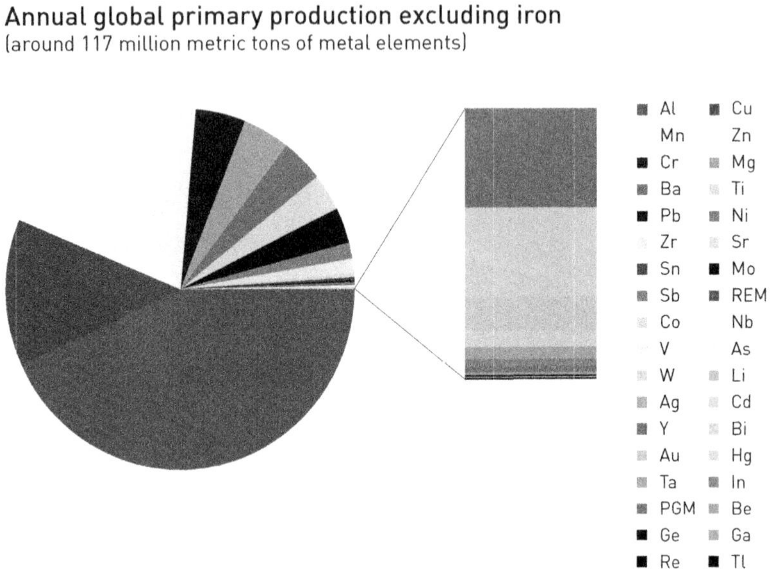

FIGURE 5
BREAKDOWN OF ANNUAL GLOBAL PRIMARY PRODUCTION OF METALS EXCLUDING IRON, BASED ON ANNEX B

On a trajectory of 'business as usual' (a modest 2% annual production growth), we will have much less than 50 years left of cheap and abundant access to metals. The production rate of metals will start to decline well in advance of the depletion of reserves as it will take exponentially more energy, materials and water input to grow or even sustain the current extraction rate of metal minerals (see Chapter 5). To sustain and increase current production rates, resources have to be extracted at more distant locations (including deep mining), at lower ore grades which require exponentially more energy to extract and with ore grades of lower quality (less beneficial composition) requiring expensive modifications of the existing metals refinery infrastructure. As already concluded in Chapter 5, metal minerals scarcity aggravates energy scarcity and vice versa, it is a reinforcing feedback loop. To make things worse, metals require huge amounts of water to concentrate

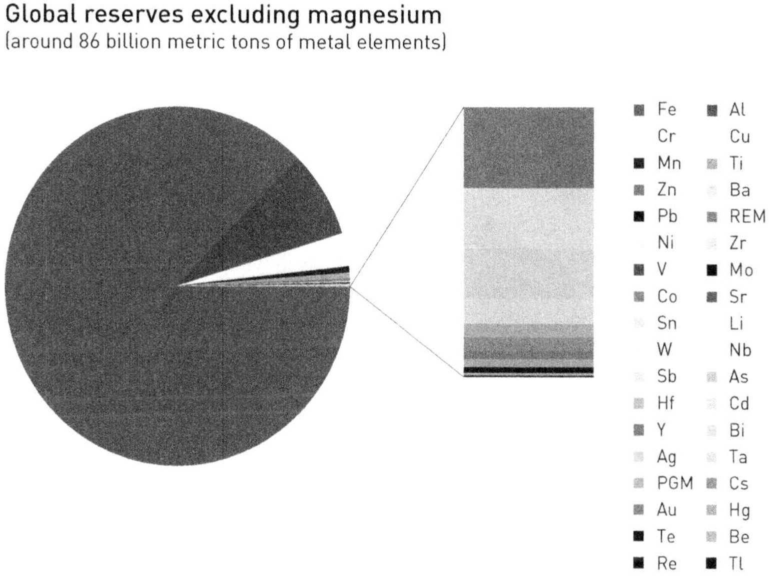

FIGURE 6

BREAKDOWN OF GLOBAL RESERVES OF METALS EXCLUDING MAGNESIUM, BASED ON ANNEX B

and also require land use (depending on the mineral being mined), in competition with other societal needs (like agriculture). As the low hanging fruit (easy accessible high ore grades) has been picked, we must now revert to harvest the less accessible and more diluted mineral deposits, requiring more input (energy, materials, water) per unit of output. Further technological progress, including efficiency gains, is still possible (see also Chapter 7), but the accumulated technological progress so far is already accounted for because it has enabled our current level of global production rates and reserves as indicated by figures 4 through 7 and Annex B.

From figures 4 through 7 and from Annex B, combined with the previous chapter, we can derive that it is highly probable that during the next few decades we will encounter serious problems mining many important metals at the desired extraction rates. Amongst them are all precious

Global reserves excluding magnesium and iron
(around 11 billion metric tons of metal elements)

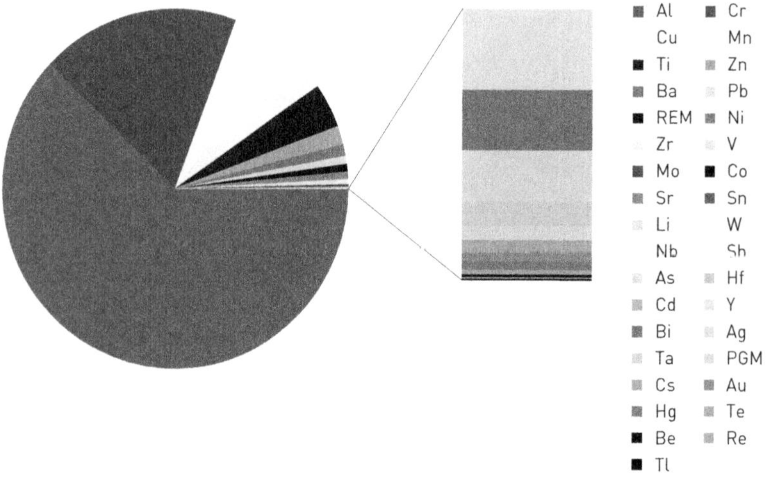

FIGURE 7

BREAKDOWN OF GLOBAL RESERVES OF METALS EXCLUDING MAGNESIUM AND IRON, BASED ON ANNEX B

metals (gold, silver and platinum-group metals), the column 3b/4b (see Annex A) minor metals (gallium, germanium, indium), part of the 'tungsten group' (tungsten, molybdenum, niobium, zirconium) and metals like zinc, tin, copper, manganese and nickel. A number of these metals are already in short supply, even if heir annual production rates continue to grow at rates well above 2% on average. Some metals, e.g., scandium, are not incorporated in Annex B by lack of data, but in most cases these metals are either abundant (like sodium, potassium and calcium) or suffer from a very low extraction rate as they are by-products (in very low concentrations) of other metal minerals. For these very-low-concentration by-product metals, independent production growth is therefore not an option, thus making an increasing role for them very difficult.

Besides the 'scarce' metals with the obvious constraint of a low R/P-ratio, we can distinguish different other 'categories' of 'scarce' metals in Annex B. First, several metals which have a high R/P-ratio (based on 2% production growth per year) suffer from relatively low absolute amounts of reserves and associated low extraction rates, effectively making them non-viable large-scale substitutes for other metals which will be in short supply. For example, it is up for debate whether lithium is a viable large-scale substitute for nickel or lanthanum in batteries for hybrid or full electric vehicles as far as land-mined lithium is concerned (it might be extracted from seawater in future[31], albeit at higher cost). Second, other metals have no acceptable substitutes for their major applications, which is of special interest for those metals which will run out relatively fast at the present course, manganese being an important example. Third, even metals with a high R/P-ratio combined with large absolute amounts of reserves and associated extraction rates, can be susceptible to future supply constraints because the high (the interesting) concentrations are mainly located in just a few regions or countries. An example is chromium which is mainly located in Kazakhstan and South Africa.

So the question of which metals are scarce, in the sense that demand will very likely exceed supply within the next few decades, will deliver a long list of metals. It is much shorter to make a list of 'abundant' metals, i.e., metals that very likely will not run into short supply within the next few decades. I will come back to this in Chapter 7 with the 'Elements of Hope', which include iron, aluminium and magnesium.

An illustration that it is usually better to be roughly right than precisely wrong (see Chapter 2) is given by the following example. A commentator on my paper[1] published at the website TheOilDrum in early 2009 claimed that I was completely wrong by putting scandium in the corner of scarce metals, because to his knowledge we could increase scandium production to around 2,000 to 2,500 tons annually from enhanced recovery from several minerals (at huge investment cost). But the real issue discussed in that paper (and here) is the overall problem of maintaining annual global primary production of all metals (excluding iron) well above a level of 117 million tons. It is not the substitution of just one metal which is being discussed in the

31 Ugo Bardi, *Extracting Minerals from Seawater: An Energy Analysis*, Sustainability (ISSN 2071-1050), 2010

paper, it's about dozens of metals becoming scarce (demand exceeding supply) more or less simultaneously. So on this scale of the problem, what exactly are those 2,000 or 2,500 tons of scandium (at what large investment cost?) going to solve?

In Chapter 3, the meaning of money was being discussed. I would like to connect this with the notion of metals scarcity by looking at the composition of coins during the last century. Of course it is not the intention of the issuer of the money that the intrinsic metal value of the coin exceeds the nominal value of the coin. The USA ceased to issue 'regular' gold coins in 1933 as did most of the world due to hoarding during the first Great Depression. Next, the USA stopped issuing silver coins in 1964 and many other countries did the same during the 1960s. In 1981, the USA stopped issuing copper cents, since then they are being made from zinc with a coating of copper. US nickels are made from copper and nickel, but discussions are ongoing to change the composition to aluminium, which is of course preferable to cheap steel (rust, stains) or plastic (toy-like money).

OVER-CONSUMPTION AND CASCADING EFFECTS

Our economic system has brought many of us prosperity and material wealth. And that's a positive thing of course. But it has also obvious downsides. In essence, in our current system everything gets monetised. Barter is nearly absent nowadays and we hardly exchange services without the involvement of money anymore. Profits should be maximised by maximising throughput. Make consumer products with such a low quality, that it is barely good enough to prevent it from remaining unsold. After it is trashed (the sooner the better) you can sell the customer a new product. And if things don't move as fast as you want, then urge on the customer using advertising and marketing. The concept of perceived obsolescence will lure the customer in prematurely replacing something which could still be used. So we end up buying stuff we don't really need or don't really use. People, for instance, have electric tools (gardening, handicraft, etc.) at home that they use maybe a few hours in a whole year. In affluent neighbourhoods, maybe all households have that same type of tool lying around without being used almost all of the time. Maybe better so, since nowadays quality standards for consumer products wouldn't permit professional intensive use anyway. It's obvious that 'unchecked' systems can grow perverse side effects. That does of course not imply that 'checked' systems can not grow perverse side effects. A rather funny example of a perverse side-effect of a system with a lot of state intervention stems from the era of production quota in the former Soviet Union. For a manufacturer of music record players, the five-year production plan prescribed by the authorities was measured in tons. The factory easily met their production goal by putting a slab of several kilograms of steel into the base of each device.

At least in the Western world, we have grown accustomed to an almost continuous and gradual increase in our standard of living. By and in itself, an increase in the standard of living is of course a good thing. But we consume more and more energy, materials and water and we even throw away a relatively large portion of our food. For many decades, every generation has had (on average) a higher standard of living than the previous one (until recently). We often feel 'entitled' to a host of products and services and will, understandably, not easily give up the things we have grown accustomed to. And it's not about necessities (need to have) but about luxuries (nice to have). Today, large parts of (Western world) populations can afford to shop in big box stores of cheap and trendy Swedish (or is it Chinese?) furniture which have appeared across countries alongside the highways at a crawling distance from the traffic jams of oversized and overweight cars so that couples who both work all week can spend their free time on Sunday searching for new furniture to replace the prehistoric lot at home which is already 3 years old. The cars which get stranded along the road on their trip to and from the big box store have nowadays more often an electronic malfunction than a classical mechanical malfunction, thanks to the electrification and computerisation of modern cars with their host of additional gadgets. A modern average car consumes much more exotic materials, consumes much more ancillary power (think electronics and airco) and is much heavier than a comparable average car 3 decades ago, also explaining the fact that the average car has no better mileage since then despite technological progress in (amongst others) engine and drive train design. A positive effect, besides all modern-day comforts, is the increased safety of cars (structural design, airbags, etc.). Other branches of big box stores sell cheap mass-produced household appliances, which have become more energy efficient but households do not consume less energy because we simply buy and use more household appliances. The list is almost endless and can be summed up by the term over-consumption.

But over-consumption has quite literally been 'fuelled' by cheap and abundant energy, largely based on cheap and abundant fossil fuels. Our collective 'abundance thinking' (see Chapter 3), together with amongst others Jevon's Paradox (or the rebound effect, see Chapter 2) and under-pricing (see Chapter 2), leads us to believe that we can continue to consume as if there were no physical limits. The notions of "the market will solve it all" and "technology will save us" are firmly ingrained in our collective mind, leading us to chase after the mirage of cheap and abundant sustainable alternative energy supplies so that we can continue our consumption pattern. The core fallacy of this believe is the assumption that we will continue to dispose of cheap and abundant metals and other materials as we indeed did for so many decades until recently. The painful fact is that the most functionally effective metals to apply in equipment to harvest energy sources which are abundant in quantity but not in concentration (sunshine, wind, etc.), are the ones in short supply for any timely realisation on the required massive scale size. And don't forget the metals for information and communication technology required to smarten and control the harvesting equipment, the storage, transportation and conversion infrastructure and the applications running on the generated electricity. These metals (see Annex A) include most of the rare earth metals (especially ^{57}La, ^{59}Pr, ^{60}Nd, ^{66}Dy, ^{65}Tb), the platinum group metals (especially ^{44}Ru, ^{45}Rh, ^{46}Pd, ^{78}Pt), a host of metals applied in battery systems (besides ^{57}La most notably ^{3}Li, ^{27}Co, ^{28}Ni, ^{30}Zn, ^{48}Cd, ^{82}Pb), a number of metals necessary for semiconductors and a variety of electronics (including ^{31}Ga, ^{32}Ge, ^{49}In, ^{47}Ag, ^{50}Sn, ^{52}Te, ^{73}Ta) and various metals required for power transmission and power conversion. Also the non-metal selenium (^{34}Se) may be in short supply for a timely realisation of alternative energy harvesting equipment on the required massive scale size. In the specific case of selenium it applies to high efficiency thin-film photovoltaic solar panels.

For all materials mentioned there may be substituting materials to try to get around the problem of scarcity. However, first and foremost, substitution of scarce materials by less scarce materials will in most cases lead to less effective processes and products, lower product performance, a loss in product characteristics, or lead to less environmentally friendly or even toxic compounds. Second, we can not just replace one material for another because the substitute candidate materials are already being used in various other applications and are often becoming in short supply themselves. Because while we are trying to make a transition towards a sustainable energy supply (using sunshine, wind, etc.), our industrial civilisation must continue and we even wish it to grow, so demand for almost all the other materials does not diminish overnight. Quite the contrary: because our industrial civilisation is still largely dependent on fossil fuels, demand for many metals and materials is continuing or increasing to satisfy demand from oil, gas and coal production, new power plants, new consumer products and more fuel-efficient transportation. And don't forget that any expansion of the renewable energy sector requires input from fossil fuels and metals as well. The largest user of metals is the entire energy sector, from exploration and production of fossil fuels and enriched uranium to energy production in power plants (including hydro and nuclear), storage, conversion and transportation of energy (gaseous, liquid, thermal and electrical). Besides the metals or applications mentioned in the previous paragraph (equipment to harvest diluted energy sources), amongst others ^{23}V, ^{27}Co, ^{28}Ni, ^{40}Zr, ^{41}Nb, ^{42}Mo and ^{74}W for steels and (super-)alloys will be in short supply within the next few decades because of the needs from our existing (and still expanding) 'normal' energy infrastructure. We will also be in short supply of barium (^{56}Ba) for barite based drilling mud used for oil and gas well drilling. An exotic metal like indium (^{49}In), required for high-efficiency photovoltaic solar panels, is already being used in flat-screen televisions and displays and in other electronic devices like mobile phones. To maximise the fuel efficiency of passenger jets, turbine engine components are alloyed with rhenium (^{75}Re) to allow for higher engine temperatures, but both the production rate and the proven reserves of rhenium are very limited. Many more examples can be given, so instead it is more pragmatic to shorten the list to 'abundant' metals, those that very likely will not run into short supply within the next few decades. As mentioned before, I will come back to this in Chapter 7 with the 'Elements of Hope', which include iron, aluminium and magnesium.

Without timely implementation of mitigation strategies (see Chapter 7), the world will soon run out of all kinds of affordable mass products and services. Already explained are the physical constraints involved in switching fast enough and at the required massive scale from current energy production to sustainable ('renewable') energy production. Those same physical constraints will prevent us from switching fast and massively to electric cars because of the massive amounts of specific metals needed for lightweight high power-density rechargeable batteries and for efficient electromagnetic motors/generators. Another striking implication will be the fading away of *cheap* mass-produced consumer electronics like mobile phones, flat screen TVs and personal computers for lack of various scarce metals. Further, a host of mass-produced products will suffer from much lower production speeds (or much increased tooling wear) during manufacturing owing to a lack of the desired metal elements (like tungsten and molybdenum) for tool steels or ceramics (tungsten carbide). Among the affected mass-produced machined products are various household appliances and all types of motorised transport (cars, trains, ships and aero structures). The lack of various metals (amongst others nickel, cobalt, copper) for high-performance steels and electromagnetic applications will affect all sectors which apply high-performance rotating equipment. Besides transportation this includes essential sectors like electric energy generation (coal/oil/gas-based and nuclear power plants, hydropower, wind power). Also the vast areas of construction work in general (housing, infrastructure) and chemical process industries will be affected. The most striking (and perhaps ironic) consequence of a shortage of affordable metals is its disastrous effect on global mining and primary production of fossil fuels and metals: these activities require huge amounts of main and ancillary equipment and consumables.

These threats to the global economy and our industrial civilisation require political, behavioural and governmental activities as well as technological breakthroughs. Of the breakthroughs, intensified recycling offers the opportunity to buy us time and innovative substitution may lead to sustainable options (see Chapter 7). But maybe the gravest scarcity of all is the scarcity of time. Reliable and detailed global information collection, analysis and assessment, problem solving, planning ahead, innovation, implementation: all these things require time. We do no longer have enough time to prepare for a painless transition, that window of opportunity has most likely passed a few decades ago. Instead, we have made some wrong choices along the way because we thought there were no physical constraints. Just think of the euphoria related to the dotcom economy at the end of the last millennium: we will simply grow our economy using cyberspace! The irony of course is that the information and communication technology infrastructure runs on scarce metals and lots of energy. And on the planet that I live on, there is not a single data carrier without a physical substrate, and these physical substrates require scarce metals and lots of energy to produce. According to several studies, the internet currently uses between 2 and 5% of the world's global electricity supply. And the internet orders have to be delivered by physical cars running on physical fuel, and so on.

Instead of the nirvana of a notional dotcom economy without physical constraints, we will now very likely face an era of intensified competition for energy, materials, water, and food within a contracting economy with less and less affordable choices. We will see cascading substitutions because so many items will get scarce within the same timeframe. Bio-fuels will not only compete with expensive fossil fuels but also with food and biomaterials (despite combinations of food, biomaterials and bio-fuels) and also with industry for water and even land use. Industry including power plants will more intensively compete with agriculture for water. Because of the enormous amount of money going around, the energy sector will compete and often beat other competing sectors for access to essential metals. An example from the first decade of this century: if the energy sector has to abandon copper or copper-containing pipes (environmental legislation), they can change to titanium pipes (which became too expensive then) or stainless steel pipes (which actually happened). Next, the surging demand for stainless steel drives up the cost of nickel in the usual chromium-nickel stainless steel pipes. And nickel gets more expensive for other concurrent reasons and for all other sectors applying nickel as well. So nickel gets substituted by molybdenum: you can substitute the relatively large amount of nickel in chromium-nickel stainless steel for a smaller amount of molybdenum. Because of the huge demand from the energy sector for molybdenum-containing stainless steel, the price of molybdenum then increased significantly (again, also for other concurrent reasons and for all other sectors using molybdenum). So the next quest is for a substitute for molybdenum. Of course the example given was more complex; many metal prices were rising anyway during that period because of the rising cost of oil. But the lesson here is: with dozens of metals becoming scarce within the same timeframe, there is not one obvious safe haven. Instead, you will likely get cascading substitutions.

Other cascading effects can entail the paradox of tempered demand for products or services which can no longer be afforded at the same scale. Supply and demand can go down simultaneously. This can have a negative impact (postponements or cancellations) on the development of for instance new (offshore) oilfields or new mining operations, resulting in lowered future production rates, possibly making future shortages worse. Like already mentioned in Chapter 5, this could result in an early peak in production and more left for future extraction, but this is of little comfort as we are interested in present production rates at affordable cost which we would like to see growing in order to grow our global economy. The tempered demand for products or services which can no longer be afforded at the same scale can also have a cascading effect on paid subscription services like mobile phone networks or internet access. These are only affordable to a mass population as long as there is a mass customer base. If part of the population can no longer afford to subscribe, the cost will have to be divided among a smaller customer base, possibly leading to higher subscription fees, which will then cause more clients to opt out, and so on. There are already developed countries with a very visible (and perceptible) degradation of their road system because the corresponding budgets of authorities go down whilst maintenance cost increase. Expect a number of similar self-reinforcing constraints to happen with other types of public or collective services. So leverage (see Chapter 4), from which we often benefited greatly during times of growth and expansion, can strongly act against us during times of contraction

The biggest threat from cascading effects is their additional leverage when calamities occur at the same time or shortly after one another. Where our complex society might be able to cope with a certain single calamity (or even a few calamities), our system likely will not be able to withstand the increased pressure of stacked calamities.

SYSTEMIC COLLAPSE

The biggest bubble in human history is digital money. As already explained in Chapter 3, financial products have no physical limits and have been allowed to grow exponentially to such a vast scale size that the value they are supposed to represent can by no means be met by future products and services on any reasonable timescale. This is gradually becoming apparent by the end of cheap and abundant energy, caused by the end of cheap and abundant fossil fuels. Economic growth has seen a dramatic slowdown or has seen negative rates in large parts of the world since the last few years before finishing this book. This economic slowdown is at least partially linked to energy scarcity, because our economy is literally being fuelled by energy. The potential assistance of this slowdown in relatively improving the situation with regard to over-consumption and associated global resource depletion and pollution is a cold comfort. We are already too far in overshoot. The cascading effect of energy scarcity resulting in the end of cheap and abundant water, cheap and abundant food and cheap and abundant metals and other materials will become more and more apparent. If the global economy resumes growth at pre-crisis levels (3% GDP growth or more) we will deplete our resources at an accelerating pace. If the global economy keeps stagnant we can no longer afford to make the necessary investments to just keep up our current global production rates for long (let alone make them grow). In both cases we will sooner or later reach peak resource production levels, to be followed by a long tail (see figure 3 in Chapter 5). That wouldn't have to be such a big problem if it wasn't for the proportional relation between GDP growth and consumption growth of energy, materials, water and food on a global scale. The prospect of even a possibility of systemic collapse, i.e., collapse of our entire global industrial system, is hard to accept for almost anyone. And if someone is willing to give it a thought, he or she sets his or her hopes on some 'graceful degradation' of the system, But without far-reaching and timely action, a stepwise descent down the stairs of 'current accomplishments', an incremental collapse, is perhaps the more likely outcome. Joseph Tainter[32] pointed out that complex systems do not self-simplify; they collapse catastrophically and are eventually replaced with much simpler systems. Dmitry Orlov[33] pointed out that "a system that evolved in conditions of continuous growth of material resources cannot shrink controllably". We face the dilemma that we can't change our current system far-reaching and fast enough without damaging or even destroying it. And we will face fierce resistance against change from anyone who benefits from the current system; their stakeholders will use complexity as a strategic argument *against* change.

---

32 Joseph A. Tainter, *The Collapse of Complex Societies*, ISBN 9780521386739, 1990

33 Dmitry Orlov, *Our Future and the End of the Oil Age: Building Resilience in a Resource-Constrained World*, Bristol Community College, University of Massachusetts, Dartmouth, March 25, 2010

So collapse as a credible possibility is being largely ignored. But it happened before to other civilisations. Jared Diamond[34] gives ample examples from our past human history. He points out that environmental degradation and depletion of natural resources are the greatest threats to civilisations. Systemic collapse won't be an event of course, but a process. At our current rate and scale-size of global resource consumption it will probably only take a few decades to unfold. Dmitry Orlov[35] distinguishes 5 stages of collapse: 1. Financial collapse, 2. Commercial collapse, 3. Political collapse, 4. Social collapse, 5. Cultural collapse. We are already in Orlov's stage 1 (since around the end of 2007) and we are probably only years away from entering Orlov's stages 2 and 3 at the time of finishing this book (2010). Can we avoid reaching the higher numbered stages? The next Chapter will explore the possibilities of a viable adaptation framework. Because all that matters now is adaptation. Or as Charles Darwin put it: "It is not the strongest species that survive, nor the most intelligent, but the ones most responsive to change".
But before we go to that next chapter, we will first briefly take a look at climate change and the increased prospects for conflict as these matters can and most likely will further complicate and aggravate our predicament with regard to global resource depletion.

CLIMATE CHANGE

As explained in Chapter 3 all models describing complex matters, including our current climate models, are incomplete and hence have large uncertainties with regard to their predictive value. So it is very difficult to accurately predict the average global temperature or the rise in sea levels in 2025, 2050 or 2100. Our scientist do however analyse past and current events and derive working mechanisms and trends with which they can estimate the probability of parameters like temperatures and sea levels lying within certain ranges in the future. But our climate system is a very complex one. A number of scientists say that around 60% of the greenhouse effect, which by the way is a benign mechanism because we wouldn't have such a human friendly climate

34 Jared Diamond, *Collapse - How Societies Choose to Collapse or Succeed*, ISBN 9780670033379, 2004
35 Dmitry Orlov, *The Five Stages of Collapse*, Energy Bulletin, November 11, 2008

otherwise, is being caused by water vapour (H_2O). The other 40% of the Earth's greenhouse effect is being caused in roughly equal shares by methane gas (CH_4) and carbon dioxide (CO_2). We seem to focus a lot on CO_2 emissions, where we might want to put more focus on H_2O and CH_4 (besides CO_2). The Amazon forest for instance pumps large amounts of water into the atmosphere but is being cut down at an alarming rate with unknown but potentially catastrophic destabilising effects on the world's climate system. The thawing of the vast tundra areas in the northern hemisphere, releasing increasingly large amounts of methane, holds equal potential for catastrophic destabilising effects on the world's climate system. As we saw in Chapter 4, these forcings, which include the ever rising levels of atmospheric CO_2, acting upon the complex adaptive system of our global climate, can result in unchecked reinforcing feedback loops that could ultimately destroy the current balance and tip our climate system into another state which is much less benign for humans.

Despite all the uncertainties and the lack of a complete understanding of our complex climate system, we do understand some of its basic mechanisms. Again, thermodynamics (see Chapter 2) plays a key role. Because warmer air can contain more moisture, a warming globe possibly tends to make dry areas drier and wet areas wetter, and possibly tends to shift the moisture distributing atmospheric circulation further away from the equator and towards the poles. Climate zones will raise in temperature and will also slowly migrate towards the poles. And as John Michael Greer strikingly explains[36], an increase in average global temperature leads to "an increase in unexpected and disruptive weather patterns on *both* ends of the temperature scale"[36]. This is because it is exergy (see Chapter 2) – the difference in energy concentration between the atmosphere and the oceans– that determines how much work a system can do, and "the work that the Earth's atmosphere does is called 'weather'"[36]. Expect our weather to increasingly "work harder".

We would need additional resources to combat and mitigate the likely consequences of climate change on essentials like fresh water availability, food production, coastal defence and heating or cooling buildings. And we learned from the previous chapters that we will

36 John Michael Greer, *An exergy crisis*, March 3, 2010, http://thearchdruidreport.blogspot.com/

dispose of less and less resources in the near future due to global resource depletion. So, climate change will make things much worse, probably at an accelerating rate.

INCREASED PROSPECTS FOR CONFLICT

History gives ample examples of shortages (water, food, energy, land) leading to conflict and all-out war. Shortages result in hardening of relations on all levels, from individuals to groups of people to city districts, regions and states. When there's enough for everyone we can afford to show compassion and be friendly and civilised, but resource depletion creates harsh competition. When things get tough, people and their higher aggregate levels up to entire states put self-preservation first and everything else is pushed aside. The last six-plus decades have been quite peaceful for large parts of the world and especially during the last few decades after the end of the Cold War (±1945 – ±1989) we have enjoyed a level of prosperity (at least in the Western world) at a scale never seen before in human history. During the last 4 decades, politicians in many countries could afford to send out their military on 'peace keeping' missions with strict 'rules of engagement' and also put efforts in civil reconstruction activities. We can expect increased resource nationalism, more corruption, increased repression, more totalitarian regimes and uncontrollable refugee streams. Many of the now prosperous countries will shake of their naïve vision that he who does no harm has no enemies and they will rediscover that if you *are* an enemy (within the eye of the beholder), you *have* an enemy (that same beholder). In the eyes of the deprived, international law and property rights may lose their legitimacy when severe shortages and unbearable inequality of distribution of the remaining resources (water, food, energy, materials) occur. We will rediscover that in life and death situations, fighting 'according to the rules' is really fighting with a handicap. War, besides resulting in human tragedies, will of course just speed up global resource depletion. War also will cause serious logistic bottlenecks, thereby aggravating the consequences of the nonlinear distribution of resources and production capacity across the globe, resulting in much earlier regional and local shortages of resources than would be the case without war.

CHAPTER 7
MANAGED AUSTERITY AND THE ELEMENTS OF HOPE

INTRODUCTION

As explained in Chapter 5, we are facing energy scarcity and energy scarcity means metals scarcity, or more general materials scarcity. Materials scarcity in turn restricts the materialisation of a new infrastructure needed to harvest diluted energy sources like sunshine and wind as alternatives to fossil fuels. Some materials are much more constrained than others, but within the holistic approach of global resource depletion and the simultaneous requirements for materials from various sectors throughout our industrial society, there is not one single solution direction that will solve materials scarcity. Important solution or adaptation directions include efficiency, recycling and substitution but a viable solution or adaptation framework should include a focus on the most abundant elements or the "Elements of Hope" together with using less (also in an absolute sense) or "Managed Austerity". We will have to deal with multiple dilemma's, we will have to decide upon many trade-offs and we will also have a host of innovation opportunities.

LEAST REGRET OPTIONS AND ADAPTABILITY

As discussed earlier in this book, there is no certainty. The world is governed by laws relating to probability, not to certainty. You cannot avoid surprises. And the world is much too complex to nicely fit in simple models suitable for reliable predictions.

So any planning for the future should be adaptive to avoid the pitfall of preparing for a specific future scenario which will never evolve. Sensible preparations should leave you with the ability to respond in a flexible way regardless of how things turn out. If you think out 150 specific scenarios, the future will evolve according to scenario number 151, which you didn't conceive of. So it's better to not lock yourself into a fixed course. Sensible preparations give you options and possibilities. Prepare for a range of futures and keep in mind that there will be surprising and unpredictable consequences of global resource depletion.

People who really understand this can be found amongst others in the military. War by its nature is unpredictable. Defence planning rests upon guesswork. There is no certainty, there are no waterproof calculations involved. It is based on identifying the issues that are most important, the notion that you will inevitably make mistakes but should try to restrict them to the minor issues and the acknowledgement that you will be surprised many times. So your plan doesn't need to be perfect as long as it covers the significant variables and is surprise-tolerant to them.

This approach resembles the notion of "no regrets" or "least regrets" options. These are options that work in both best case and worst case scenarios. They should pass the test of answering "no" to both the question "if everything goes well, do I regret my preparation?" and the question "if everything goes bad, do I regret my preparation?" The least regrets options are the options with intrinsic value and therefore can be used to mobilise support by putting emphasis on the most acceptable reasons. Politicians will probably use (or are already using) this to avoid the need to uncover other inconvenient or disputed reasons. In this way policy measures to mitigate or soften the effects of resource depletion can gain support and can be legitimised using for instance climate change (fossil fuels scarcity) or environmental or health concerns (metals scarcity) as *main* arguments.

For the EU, which as a whole is energy-poor and materials-poor, a least regrets strategy could include, for instance, a far-reaching focus on energy-extensive and materials-extensive products and services. Let's presume that collective interest wins from competitive disagreement (as it should) and the EU implements such a strategy. If things really go bad (see Chapter 6), the EU will have made itself less dependent upon other parts of the world for scarce and expensive energy and materials and has prepared itself to a certain extent for a resource-constrained world. If things go really well and global resource depletion doesn't bite (yet), the EU will have realised a meaningful step towards a more sustainable civilisation with reduced greenhouse gas emissions, reduced pollution levels, diminished negative impact on biodiversity and new industrial activities able to compete with more energy- and materials-intensive activities from elsewhere. In the ideal case, eventually the world as a whole would make a transition towards an energy- and materials-extensive economy.

MANAGED AUSTERITY

As we saw in Chapter 2, a system left on its own strives for minimal free energy content. So, if we want to have a lifestyle better than that of a hunter-gatherer we have no choice but to interfere, as we did since ancient times: agriculture, crafts and trade, more complex shapes of governance, and so on. But now that our current level of interference is clearly unsustainable with its associated energy throughput and complexity, we have no other choice than to 'retreat and consolidate' to retain as much safety and prosperity as possible and not loose everything. The likely alternative is to briefly extend our overshoot with subsequent systemic collapse (see Chapter 6).

The big dilemma of course is changing our current system into a state of non-overshoot without promptly triggering collapse. The intention is to have a gradual decline, a 'graceful degradation', to a state as high as possible above that of some pre-industrial era. But consumption growth is firmly built into our industrial civilisation, making it quite likely that, without determined intervention, we will continue speeding towards the end of the track and accelerate instead of hitting the brakes. The end result will likely be large-scale contraction and collapse (see Chapter 6), possibly within one generation. Also World War 3 cannot be ruled out as a possible result.

So basically we cannot avoid having some kind of 'Managed Austerity', i.e., lower rates and lower absolute levels of resource consumption. It's either that or we'll stubbornly dig our own grave prematurely.

Of course, this is not what we want to hear, so instead we cling to the false hope that technology somehow will 'solve' our 'problem'. However, 'solutions' to 'problems' all too often are encouraging the original problems to grow as we saw in Chapter 3. This is the main reason why we need Managed Austerity. Let me explain using the case of metals scarcity. The production rate of metals is, besides by capital investment, primarily limited by the constraints mentioned below and the available facts make it likely that we are nearing the limits for most metals within years or decades, not centuries (see Chapter 6). Technology is expected by many (see also Chapter 3) to 'escape' the constraints summarised in simplified form in Table 1. Ironically though, new technology on its own, without some form of control or restraint, will advance the collision between unlimited growth and limited supply because it enables consumption growth.

TABLE 1
TECHNOLOGY FACILITATES SUSTAINED CONSUMPTION GROWTH

Constraints:	Main technological reactions:
Geological: Availability of accessible high-grade ore bodies	Ocean floor mining *results in higher resource consumption (if feasible at all at large-scale size)*
Physical: Required input of resources (energy, materials, water) for mining and conversion of ores into metals	Efficiency *results in higher resource consumption*
Geopolitical and geo-economic: Concentration of high-grade ore bodies in few countries or with few owners	Substitution *results in higher resource consumption*
Environmental: Habitat destruction, pollution of soil, water and air	Recycling *results in higher resource consumption*

Key to meaningful and timely adaptation to global resource depletion is to cure from our *rate addiction*. We want more; we want if fast; we want it now. We are not stressed because we don't have time; we don't have time because we are stressed. And our IQ gets a dip when we are in a rush. Our hurry is causing us to dig our hole deeper at maximum speed. We know from the laws of thermodynamics that only infinitely slow processes have virtually no losses. The faster we want things to happen, the more resources we waste using evermore complex and wasteful technology. We have a lot to gain from scaling back to more moderate rates of resource consumption. We need technology, but we have to apply it in an optimal way, not in a maximising manner. Nature has inspired humankind throughout history for its beauty and ingenuity. Where humans tend to maximise, nature has learned to optimise within its vast system boundaries. Some humans tend to measure and compare the efficiency of a leaf to that of a high-efficiency photovoltaic solar cell made from semi-conducting materials. Those people miss the point that a leaf provides much more services than just converting solar radiation into another form of (partially stored) energy. We should be striving to match nature instead of striving to surpass her. Humans have nanotechnology (the term is often used for things on a micro-scale as well), whereas nature is capable of 'engineering' on a molecular scale with far less energy and auxiliary materials (and health issues). So again, the question is not whether we can beat nature, but if we can match nature.

We also have to use less in an absolute sense. If we do this voluntarily, this will prevent us from being forced into much lower levels of material security by systemic collapse. We have to contract and simplify and if we utilise our ingenuity and creativity and act timely and accordingly, it's quite possible that we could realise this without detrimental effects. Managed Austerity within this context can be defined as the combination of moderate rates (production, conversion, consumption, etc.) and using less in an absolute sense. "Using less" itself has to occur at a moderate rate to avoid collapse of our current economic system.

Managed Austerity is austerity we can manage. Managed Austerity does not necessarily mean that we all have to live like impoverished people with few possessions and no luxuries at all. Musical instruments for instance are not necessary for survival but nevertheless are 'useful luxuries'. The largest gain lies in abandoning the many 'non-useful' luxuries, starting with the most excessive ones. Food is of course a necessity, but out-of-season fruits and vegetables flown in from another continent obviously aren't. It's not about the fruit and vegetables from other continents (they can be partly shipped), it's about the waste of precious energy (unnecessary air freight and associated logistics) better utilised elsewhere. Other examples are cars for public streets able to drive more than two times the speed limit on a highway or able to accelerate like a rocket once the traffic light turns green. More importantly: what about the mass production of low-quality consumer goods made barely good enough to avoid them from remaining unsold and usually ending up in a landfill only months or a few years at best after buying?. In the next paragraph a general framework to adaptively cope with resource scarcity will be discussed with the knowledge from the current paragraph that all these measures do not work properly (from an holistic point of view) without the additional control or restraint by Managed Austerity.

THE ELEMENTS OF HOPE

There are six adaptation directions to diminish our dependence on scarce resources, most notably energy and metals (or materials in general): using less, longer product lifetime, more intense recycling, substitution with less scarce metals (or other materials), a new product or process design philosophy and adapted inventory management. These are 'least regret' adaptation directions because they are meaningful no matter how the future is going to unfold; they are solution directions with intrinsic value. The six directions and their various combinations form a consistent framework and the various (combined) parts of that framework are known under many different names, e.g., the catchy notion of 'dematerialisation' (like film tapes which have been replaced by digital photos). Part of the adaptation framework is facilitated by reducing complexity (see Chapter 4) in order to enhance quality and diminish waste. Complexity reduction in turn is facilitated by selectivity and simplicity as explained by the Pareto principle (see Chapter 2). The adaptation framework includes a number of balancing feedback loops to counteract the reinforcing feedback loop of energy scarcity and metals scarcity discussed in Chapter 5.

USE LESS

Of course the 'greenest' energy is the energy you don't use. And the 'greenest' materials are the ones you leave in the Earth's crust. The core issue with regard to using fewer resources is to slow down and end global population growth and to slow down and diminish overall consumption rates by the population which is already present. Of course these are largely political taboos. More practically, using less can be combined with one or more of the other adaptation directions to gain higher efficiencies and to enable 'dematerialisation'. Dematerialisation often boils down to replacing a large quantity of one material by a much smaller quantity of another material (in fact a combination of using less, new design and substitution). An example is the replacement of film tapes by digital photography. We have largely abandoned traditional cameras using film tapes so we don't need all those silver containing negatives to be developed in a bath of chemicals. But please keep in mind that dematerialisation still involves materials: you still need physical substrates to store the digital information and people still like to make prints. We use exotic materials in the cameras, the digital picture frames and the memory chips. In order to have benefit from efficiency gains and dematerialisation solutions, care should be taken to avoid the pitfall of Jevon's paradox, e.g., more kilometres per year travelled in a more fuel efficient car or making more photo prints or using digital photo frames running on batteries instead of framing a print.

A particular pitfall with dematerialisation is the danger of solving or alleviating the problem of scarcity of specific rare metals by using very thin layers or by using nanotechnology, whereas the associated production processes are consuming large amounts of energy[37] and consumables. From an holistic (whole system) point of view we would risk making the classic mistake of only solving part of the problem, i.e., specific metal scarcities, whilst aggravating the core problem, i.e., energy scarcity. Moreover, thin layer technology and nanotechnology are complicating the much needed efforts towards more intense recycling.

More a behavioural change than a 'technical' solution is to focus on essential goods and services, i.e., basic necessities as opposed to short-lived luxuries. People should (re-)learn the difference between inconvenience or discomfort on the one hand and real (even life-threatening) problems on the other hand. We have to make a distinction between luxuries or 'nice to have' and necessities or 'need to have'. The first thing to abandon is planned and perceived obsolescence, especially applicable to various consumer products. Planned obsolescence is the strategy to make products with such a poor quality that the rate of consumption is maximised whilst the quality is marginally sufficient to prevent the products from remaining unsold. Perceived obsolescence is making people believe (using marketing) that they should buy new stuff

37 Gutowski et al., *Thermodynamic analysis of resources used in manufacturing processes*, Environ Sci Technol., 2009

to prevent them from being seen as morons or to be more successful etc. Marketing and advertising should be used in future instead to support implementation of Managed Austerity and the Elements of Hope.
Using less has the intrinsic value of acting as a delay in a complex adaptive system (see Chapter 4), giving balancing feedback loops (such as recycling and substitution) or mitigating or softening effects (such as starting up new primary production) time to gather steam. Using less is also slowing the gain around the reinforcing loop of consumption growth (see Chapter 4), thereby again giving the balancing loops time to function. As already mentioned, part of 'using less' is to achieve better efficiencies, including using existing systems with maximum efficiency. This is facilitated by global transparency of information and experience and includes sharing of technology (see Chapter 4).

LONGER LIFE

Longer product lifetime is facilitated by reducing complexity (transparent technology is helpful here) and by design dedicated to easy maintenance and repair. A slide rule is a wonderful example of transparent technology: it's much easier to comprehend how calculations are being performed than using a calculator (do you know how the microchip works in detail?). Moreover, the calculator needs batteries and its lifetime is much less than a durable (non-plastic) slide rule. Cars used to have space left under the hood (sober designs), amongst others making it easy to change a light. Many modern cars (often luxury designs compared to the past) have to be taken to a workshop for a simple task such as changing a light. The light is just a simple example but the same 'logic' shortens the economic life of complex designs. The breakeven point after which it is less costly to buy a new product is reached rather quickly in our current system.
A strong incentive for longer lifetimes (as well as better efficiency and more intense recycling) can be the emerging business concept of selling services rather than products. A manufacturer of lighting hardware for instance makes a deal with the hirer of a large office building to deliver a certain amount of light per room for a number of years at certain cost. The manufacturer may even take responsibility for the electricity consumed by the lighting. It is now in the direct interest of the manufacturer to install lighting hardware with a long lifetime, with minimal need for maintenance and repair and with

minimal energy consumption. When the manufacturer has developed a new lighting hardware generation which is even more energy efficient, he can decide that it's more profitable to change the lighting hardware and recycle or reuse the old hardware somewhere else, as long as he delivers the agreed upon amount of light per room. A similar example exists in commercial aviation, where some airline operators prefer to rent their gas turbine engines, because they don't want to own them and bear the risk of unforeseen maintenance. It speaks for itself that it is in the direct self-interest of the gas turbine engine manufacturer to install hardware with maximum reliability and lifetime.
Like 'using less', longer product lifetimes act as a delay and slows down the gain around the reinforcing loop of consumption growth, giving balancing feedback loops or mitigating effects time to kick in. And sharing of technology and 'best practices' on longer life on a global scale wouldn't hurt either.

RECYCLING
Recycling the current and constantly growing inventory of metals (and other materials) in use in various compounds and products is the obvious choice in order to buy time and avoid or diminish short- to medium-term supply gaps. Although recycling is nothing new, generally the intensity could be further enhanced. We should keep in mind though that recycling has inherent limits, because even 100% recycling (which is virtually impossible) does not account for annual global demand growth. At the present course we need to continue to expand the amount of metals in use in order to satisfy demand from developing countries like China and India whose vast populations wish to acquire a material wealth comparable with the standard of living of the industrialised Western world. Furthermore, recycling also costs lots of energy (progressively more with more intense recycling) and many compounds and products inherently dilute significant parts of their metal constituents back into the environment owing to their nature and use. So even with intense recycling, we will need a continued massive primary production to continue our present collective course.

Cradle to cradle (C2C, see also Chapter 3) is not the single solution to metals (or materials) scarcity either. C2C could be summarised as creating flows rather than deplete stocks, which is a good thing of

course. If stock depletion would be depicted as a straight line, C2C could be depicted as a curl. The biggest fallacy of C2C is the absence of some form of control or restraint on resource consumption, even if most of those resources would be waste. In this way, the C2C concept can be depicted as a tree full of branches loaded with curly leafs. The leafs have the proper shape (curly, i.e., a circular process), but due to the absence of some form of consumption constraint in the C2C approach, we end up with an ever growing tree (using lots of energy, nutrients and water), sprouting and growing branches with evermore curly leaves. So recycling, like all other parts of the adaptation framework, by and in itself it is not enough, it should be accompanied by Managed Austerity. But recycling is a strong and valuable adaptation strategy, it's a strong balancing feedback loop. It's also an important business opportunity in a resource constrained world. On a global scale, the more abundant metals are recycled at relatively high degrees, whereas the scarce metals are mostly recycled at very low rates or not at all. This offers great prospects for future improvements. Moreover, new recycling technologies can have enhanced leverage because they will not only benefit recycling but will also enable more efficient extraction and concentration of certain metals in the primary production chain.
Again, sharing of 'best practices' and technology, even if only restricted to the proper generic methods, wouldn't hurt.

SUBSTITUTION

Substitution (like recycling) can act as a strong balancing feedback loop and can act as a delay by slowing down demand for the resource which is being substituted (or recycled). It is self-evident that, at our current level of technology, substitution of scarce metals by less scarce metals (or other materials) for major applications will lead to less effective processes and products, lower product performance, a loss in product characteristics, or will lead to less environmentally friendly or even toxic compounds. An important and very challenging task is therefore to realise the desired functionalities of such products with less scarce materials and to develop processes for production of these products at an economic scale. A particularly powerful solution direction is the substitution of scarce metal elements by the most abundant elements, the so-called Elements of Hope which you could also name 'elements of abundance'. These are the most abundant elements available to mankind and can be extracted from the earth's crust, from the oceans and from the atmosphere. They constitute both metal and non-metal elements. We can look at the remaining producible global reserves of metals as a toolbox for current and future generations, see figure 8.

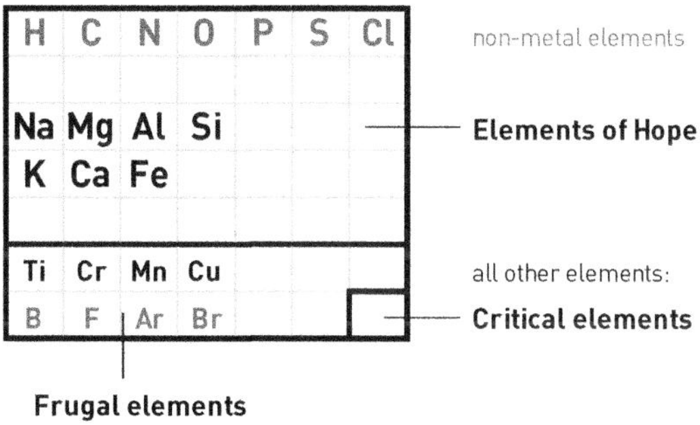

FIGURE 8
THE TOOLBOX CONTAINING THE ELEMENTS OF HOPE, THE FRUGAL ELEMENTS AND THE CRITICAL ELEMENTS; Si (SILICON) IS A METALLOID; THE RED (OR GREY) ELEMENTS ARE NON-METAL ELEMENTS

The largest part of the toolbox, based upon the findings of Chapters 5 and 6, is reserved for the Elements of Hope. Another part of the toolbox is reserved for less abundant but still plentiful building blocks, the *Frugal elements*. These elements should be used predominantly for those applications for which there is no substitute with current technology (example: chromium for stainless steel). In this way their remaining reserves will last longer (most notably copper and manganese). Finally a small corner of the toolbox is reserved for all other metal (and non-metal) elements combined, the *Critical elements*, which should be saved for essential applications where substitution with less scarce elements is not possible.
Not coincidentally, all macronutrients of nature (all flora and fauna including the human body) are found among the Elements of Hope (see figure 9): living nature either uses these elements (metabolism, building blocks) or has evolved tolerance to these elements in their abundant natural forms. Substitution based on the Elements of Hope can therefore be inherently environmentally friendly, they lack any heavy metal.

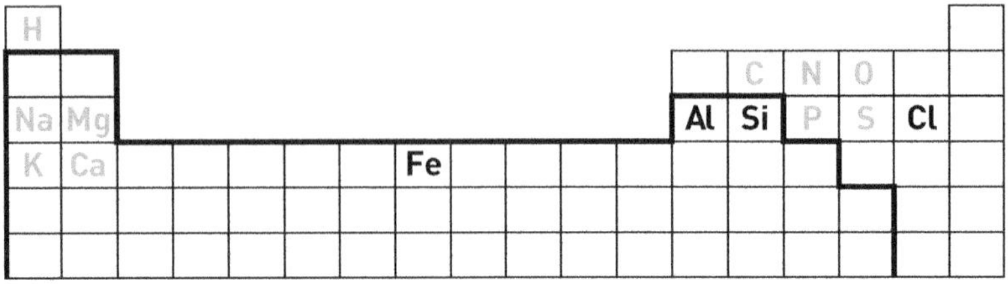

FIGURE 9

THE ELEMENTS OF HOPE; THE GREEN (OR GREY) ELEMENTS ARE MACRONUTRIENTS, THE ELEMENTS WITHIN THE THICKENED SECTION ARE METALS, Si (SILICON) IS A METALLOID

Hydrocarbons for production of materials (including plastics) could be extracted progressively more from biomass, albeit at a much lower extraction rate than from concentrated (fossilised) biomass (oil, gas and coal). The feasible scale size of biomass materials production is limited by amongst others water availability, competition with food and energy (bio-fuels), availability of arable land and the feasibility of proper replenishment of soil nutrients to prevent biomass production becoming 'mining the soil'.

The term 'decarbonisation' within the context of emissions control is a rather poor choice, because carbon as such is not the problem. Quite the opposite, carbon is one of the Elements of Hope.

It should be noted that our future choice for substitution will become more and more limited because of depletion of producible reserves (see figure 3 in Chapter 5). This could be a strong incentive to aggressively build out a new alternative energy infrastructure with the best performing metals, now that these metals or their associated products are still affordable and available for energy production purposes. Examples are large wind turbines using neodymium and dysprosium containing permanent magnets and high-efficiency thin-film photovoltaic solar panels using gallium, indium or tellurium. Measures should then be taken to ensure that the embedded strategic and scarce metals remain available for recycling and reuse within the country or region of application at the end of life of the wind turbines and solar panels.

Once again, sharing of 'best practices' and technology, even if only restricted to the proper generic methods, wouldn't hurt.

PRODUCT AND PROCESS DESIGN

As we saw in Chapter 4, changing the rules is a powerful lever. Industrial designers for instance can have a large impact by designing products radically differently by consistently using the Elements of Hope wherever possible. Products should be fit for extreme (overall around 90%) recycling, including design for disassembly which could benefit production efficiency as well and including design for material combinations which facilitate easy recycling (certain metals shouldn't be combined). Products should also be fit for a long life based on high quality and intentional reparability instead of planned (or perceived) obsolescence. And process engineers can have a large impact by ruthlessly redesigning and applying process routes that save as much energy as possible. Of course such changes will lead to certain compromises, for instance reduced product specifications, but the law of the diminishing returns works in our favour if we take the opposite direction. If it took, for instance, twice as much energy per product to increase certain product characteristics by 10%, then going back and reducing the energy input by 50% still will result in retaining 90% of the product characteristics. In a resource constrained world, there will be plenty of business opportunities for companies offering affordable alternatives to expensive and less available products that utilise the maximum achievable to the fullest extent. Simplification can offer a lot of leverage.

Not coincidentally, important material technology breakthroughs from a less energy-intensive and less material-intensive past may point towards material developments which are suitable for a less abundant future. We can learn a lot from history and prior knowledge. An example is given by low alloy steels developed during the 1930s which are viable only recently for reliable and quality-consistent mass production owing to modern (computer controlled) process technology. So we can benefit from "re-search" in its most literal meaning.

The practice of replacement of components or upgrades of components (e.g., ship and airplane engines or engine parts) could be taken to extremes and thus include smaller items like household appliances and other consumer products by making labour much less expensive relative to materials. Consumer products should be designed or redesigned accordingly. It will then again (like decades ago) make more sense to repair things or replace components instead of discarding complete subassemblies or complete products like we often do today. But making human labour much less expensive than materials requires us to change the way money flows (see the paragraph on implementation).

BUFFERS

With the arrival of the age of scarcity industrialism[38] a lot of common wisdom arisen in times of plenty might need some revision. Our current system won't function in reverse. Such 'abundance thinking' common wisdom might be the 'just in time' philosophy to minimise cost and maximise turnaround rates and profits. In times of scarcity with associated high price volatility it might be profitable again to have buffers to be able to cope with supply disruptions and to have the luxury to temporarily stop purchasing raw materials and products at unfavourable high cost during price peaks. On a larger scale, national stockpiles of strategic metals, like the USA used to have during the Cold War era, might be reinstated. In fact, countries like Russia, China, Japan and South Korea have national policies on strategic metals, crucial to their (exporting) industries. Ideally, strategic stockpiles will be modest in size to prevent large market disruptions (except when one wishes to set the prices and terms of course) and to prevent them from costing too much money. There is still a lot of opposition in large parts of the Western world against the notion of government intervention in the form of buffers and stockpiles of strategic materials. But don't we already do this with oil, the three-months-minimum stockpile for OECD countries? Where will our future imports come from and at what price? If things go wrong, buffers will stabilise the system by preventing (to a certain extent) supply disruptions and will enable rationing for a longer duration, thereby buying precious time for other measures to be taken. And if all things go well and scarcity will not bite (yet), the least-regret benefit from national (or international) stockpiles of strategic materials will be a certain level of supply chain confidence.

38 The term "scarcity industrialism" has been coined by John Michael Greer, see his extensive work at http://thearchdruidreport.blogspot.com

IMPLEMENTATION

The central issue of global resource depletion, comprising amongst others the interlinked scarcities of energy, materials, water and food, is energy scarcity. Energy scarcity is mainly driven by the depletion of the high-quality and easy accessible fossil fuels (regular oil, regular natural gas and hard coal), resulting in a maximum production rate of *energy per capita* (see Chapter 5) staying more or less flat for 3 decades now. It is highly likely that the energy per capita production will start to decline before the 2020s. All other energy sources combined, suffering from lower net energy densities and the reinforcing feedback loop of requiring energy-intensive metals and other materials (and water) for materialising and operating the corresponding infrastructure, cannot bridge the gap. Because of the urgency and severity of the threat of global resource depletion, we should drastically reduce our energy use per capita. Maybe we should be lucky if we could maintain an energy per capita use (global mean value) by 2030 similar to the mid-1950s which is shortly after abandoning rationing in Europe after World War II. In order to realise this, the six adaptation directions (the Elements of Hope) and their combinations have to be pursued. But these adaptation directions are not enough, they have to be combined with Managed Austerity to prevent the adaptation framework from just facilitating continued (temporary) consumption growth, resulting in subsequent systemic collapse.

Because energy scarcity is the core issue with regard to global resource depletion, it's worth to consider if 'managing' austerity by *primarily taxing the use of fossil fuels* (without raising taxes overall) and strongly reduce taxing everything else, including human labour[39], would be an effective balancing feedback loop. Of course reality might be much too complex for such a simple, although theoretically effective, measure to function without unintended consequences. But the world disposes of an army of subject matter experts who have proven to be clever, inventive and creative in solving financial "issues" (although not always in the common interest), so they may focus their efforts in future on combating the disadvantages of taxation interference with clever solutions (in the common interest). If we would implement such a notional measure

39 Femke Groothuis of Ex'tent (**www.extent.nl**) has given a presentation on *Value Extracted Tax* during the Bioneers Global conference in Driebergen, The Netherlands on June 1, 2010

strictly and consistently, on a large enough scale (e.g., a block like the EU), with as few compromises as possible, we would quite soon be making a transition to a much different society, largely pursuing the desired adaptation framework of the Elements of Hope. If fossil fuels and consequently electricity generated from fossil fuels become very expensive, there will be a very strong incentive to save energy: higher efficiencies, improved processes, lower rates, lower energy intensities, etc. There will also be a very strong incentive to generate energy using alternatives such as solar energy, wind energy, geothermal energy etc. and maybe also nuclear energy. It could function like the German feed-in tariff for renewable energy "on steroids". Because of the energy intensity of the metals and materials required to materialise amongst others solar panels and wind turbines, there will be equally strong incentives to change and innovate their designs, to substitute materials (most notably to the Elements of Hope of figure 8) and to recycle materials. Metals themselves will be concentrated using more energy efficient processes, such as heap leaching, which is already in use for some tailings and low ore grades where land is available and cheap (e.g., desert-like areas). Low-rate methods like heap-leaching have low production rates (unless utilised massively in parallel) but energy use is also low compared to high-rate processes. Processes operating at lower temperatures benefit from lower exergy losses (see Chapter 2) and generally require less exotic, high embedded energy materials. Water will cease to be under-priced and consequently be wasted much less as the high cost of fossil fuels works its way through the embedded fossil fuel energy in the required equipment and infrastructure (pumps, pipes, sewage treatment, desalination plants etc.) and in the electric power on which all this equipment runs (including the embedded fossil fuel energy in photovoltaic solar panels and wind turbines). Exotic metals will be even more costly than now, and although it's not forbidden to use them (and in some cases even necessary), the incentive will be very strong to do without them, even it this means accepting some loss in performance. Photovoltaic solar panels for affordable large-scale applications for instance may be based on amorphous silicon or organic materials and strong permanent magnets for instance may be based for an increasing number of applications on manganese-aluminium alloys. Many products and many types of infrastructure which are energy-intensive to produce will get longer lifetimes. Quality will be more important than quantity. People will, much more than now, base

their purchase decisions on whether they really need it and whether it is durable, versatile and has a good quality relative to its cost. Reuse and repair will make a strong comeback and waste streams will be largely diminished by this and by longer lifetimes and more intense recycling. Nonmaterial services will be more affordable because of the much lower tax on human labour. Available money and resources will be invested much more towards creating new jobs and much less towards automation which renders (at this moment expensive) people obsolete. We have an abundance op people (a euphemism for overpopulation), so we'd better focus more on people and less on things.
Of course people need help, they need incentives to make the necessary changes. Changes in the way money flows won't happen overnight. People want to hear 'solutions' without making any real sacrifices and politicians better follow suit to get elected. We get distracted from the main issues. For instance, lots of people think they do a good job fighting climate change (or fossil fuel dependence) by changing a few light bulbs for energy saving lamps, whereas most energy consumption from lighting stems from office blocks, commercial real estate, industry, public infrastructure including the entire road system, etc. The terms 'green' and 'sustainable' are often abused to legitimise 'business as usual', i.e., faking feeling passionate for good stewardship of our planet while being focussed on maximising profit. That does not mean of course that sustainability and profitability would be mutually exclusive. But we should be able to make a distinction between the black, the white and the various grey shades here. If we don't want to be fooled, then we have to stop fooling ourselves. And if we want to stop fooling ourselves, we have to stop fooling others. We need genuine and sincere leaders. We need leaders who inspire and motivate us, so they have to be great communicators as well. It takes time for people to accept new things, but this will speed up as more and more people get convinced. Our leaders should be able to 'sell' us the message of the need for intervention by convincing us that all five ingredients necessary to close a deal are available: necessity, money (made available), urgency, desire and trust. We should also be aware of the pitfall of 'paralysis through analysis': numerous people and organisations worldwide are gathering and analysing information on our predicament of global resource depletion, they are trying to quantify the problem as complete and reliable as possible. This is of course a good thing to do and we should proceed with this unabated, but we have to accept uncertainty

and should not get lost in striving for complete information, which is impossible anyway. It is not enough to register what is happening (we already have an abundance of reports), we have to concurrently act upon the information and insights gathered. So we need leadership that brings together all available information, sets up a time schedule and acts accordingly. It's up for debate whether democratic political systems as they are organised today are perfectly fit for this task, especially if people are being represented by many factions. The need to reach consensus and make compromises is not very helpful here because, as the saying goes, commissions tend to agree on the lowest common denominator. This wasn't such a problem in the past decades of plenty, but this will probably act against us in times of scarcity, especially because of the scarcity in time to act.

Awareness at the level of business and government can be raised by approaching metals scarcity (and scarcity in general) as a matter of risk assessment, including its positive flipside of opportunities. Business opportunities lie amongst others in smart combinations of processes and in redefining business models. Examples range from the combination of a geothermal plant with the co-production of metals from the hot groundwater or the combination of an oil-fired power plant with the co-production of heavy metals from the combustion slag to the primary production of magnesium from seawater using excess electric power from wind and solar energy (as an alternative to battery storage). Risk assessment has to strike a balance between the extremes of fear (called sometimes Fake Evidence Appearing Real) and blind, unfounded optimism. Many leaders only want to deal with problems after they materialise. Leaders in more centralised economies such as China may have an advantage in this respect: they can focus more on long term prospects instead of the next election circus, they need much less compromising and they can act where other economies keep talking. Democracies work just fine as long as there is plenty for all, but wartime-like efforts call for triage-like decision making rather than working groups and endless debate. It is not unthinkable that global resource scarcity will lead to more centralised regimes, the Western world included. People focus more on what they don't want to loose than on what they want to gain (loss aversion). Part of the resistance against the concept of the adaptation directions mentioned in this Chapter and especially against Managed Austerity can be explained by the fear for loosing wealth and freedom if we have to go by using less and if this is accompanied by increased government intervention. The irony of course is the high risk of loosing much more wealth and freedom during a "default" scenario, i.e., we go on as if global resource depletion is not alarming but only alarmist. The default scenario holds the prospect of extreme forms of government intervention at a later stage as the unfolding uninhibited collapse triggers desperate measures.

One of the main drawbacks of outsourcing manufacturing away from industrialised Western countries during the last many decades was the disappearance of the associated basic knowledge and experience. It has been underestimated how much implicit knowledge the production level contains. More importantly, the engineering and research experts left behind in the now production-facility-scarce countries started running short of sufficient new ideas because of the loss of this knowledge base. But realisation of the proposed adaptation framework requires people's ingenuity and creativity, which partly feeds on the level where the action is. Lost manufacturing industries may be hard to re-establish in countries which now largely depend on services, but we can establish new industrial activities fitting within the framework of the Elements of Hope. Actual implementation of this framework needs a biodiversity of ideas. If we don't think out of the box, the box may become our coffin. Of course we have to keep our two feet solidly grounded and should not get carried away with the whole idea of some innovation utopia, the danger of well-intended actions having unintended consequences is omnipresent and always lies on the lookout to strike. We can for example benefit greatly from information and communication technology applications in and between products and processes to improve efficiencies and to facilitate amongst others longer lifetimes and substitution with alternative materials. But much like technology can't beat the laws of thermodynamics, automation and artificial intelligence can't beat human stupidity. And sometimes an inventor is someone who has thought of something new that doesn't function in a way it didn't malfunction before. So again, we have to remain selective (see the Pareto principle, Chapter 2). The nice thing about ingenuity and creativity is that they are not limited in the way that energy, materials, water and food are limited. So we should make a faster transition to the so-called *down-to-earth innovation economy* where creativity and common sense are the fourth and fifth factor, besides capital, labour and knowledge.

CHAPTER 8
CONCLUSIONS

Global resource depletion is real and urgent. Proper and interconnected solutions take time to develop and implement. Let us prevent going back to the order of the day after reading and processing this information and let us prevent thinking this is something to worry about later.

In essence, we now can choose between damage control (triage-like decisive action) and inaction (making no real choices) leading to systemic collapse. It's far too late for some kind of 'painless' transition. We should focus on meaningful solution directions to retain as much basic safety and prosperity as possible. We are in overshoot. We have to retreat and consolidate. We have to hand in part of our achievements and 'entitlements' to prevent us from losing almost everything.

Collapse of our current complex system is not an event with a clear defined date but a yearlong process, resulting in many people refusing to acknowledge our predicament. Don't be fooled by the trap of 'linear thinking', leading to extrapolation of the latest snapshot of unfolding events (positive or negative). The course of events will change its pace and will also change its direction from time to time. But the overall direction of our collective track is quite clear for those of us able and willing to see and understand.

Inertia is not just a law in physics; vested interests will fiercely resist change. The 'truth' will not so much be 'silenced' but will be drowned in a cacophony of selective partial truths and omnipresent misleading information. People usually overestimate threats which they can imagine quite well and underestimate abstract threats like global resource depletion. Denial will be comprehensive but will fail as events unfold. Expect revelations of overstatements of 'economically viable mineral reserves' in the longer run, preceded by abundance thinking inspired upward revisions and 'creative accounting' doing just the opposite in the coming years. Expect plenty of 'news' (like mineral deposits in Afghanistan) to distract attention from the real main issues. Security of supply is going to be more important than prices. Business and government are soon going to find out that it is naïve or outright dangerous to rely too much on all the paperwork being right. Sooner or later, rationing of certain goods and services will simply become unavoidable, probably disguised at first as being more expensive or as being more difficult to get access to. But there's no reason to panic, which is a good thing because panic tends to lead to responses reminding of the dark hospital joke involving a successful operation at the expense of a dead patient. Meaningful action can be guided by the adaptation strategies which are discussed in Chapter 7. They offer the *Elements of Hope* to cope with global resource depletion and which have both a metaphorical meaning and a literal meaning of applying living nature's building blocks. These are: using less, longer product lifetime, more intense recycling, substitution with less scarce metals (or other materials), a new product or process design philosophy and adapted inventory management.

But above all, key to meaningful and timely adaptation to global resource depletion is to cure from our *rate addiction*. Our hurry is causing us to dig our hole deeper at maximum speed. The faster we want things to happen, the more resources we waste using evermore complex and wasteful technology. We need technology, but we have to apply it in an optimal way, not in a maximising manner. We also have to use less in an absolute sense. If we do so voluntarily, this will prevent us from being forced into much lower levels of material security by systemic collapse. The combination of moderate rates and using less can be summarised by the term *Managed Austerity*. "Using less" itself has to occur at a moderate rate to avoid collapse of our current economic system.

So, we have to cure from our rate addiction and from our addiction to material wealth, especially energy-intensive products and services. The virulence of an addiction is the fact that it never fully satisfies us so that we continually seek more of it[40]. What we really want, besides basic physical necessities as clean air, healthy food, potable water and shelter from heat or cold, is nonmaterial wealth: love, friendship, a feel-

40 Carlton K. Erickson, *The Science of Addiction – From Neurobiology to Treatment*, ISBN 9780393704631, 2007

ing of belonging and connection, a sense of purpose, consciousness. In short: real wealth.

The required holistic approach to our predicament of exponential growth and over-consumption requires involvement of various disciplines. An effective medicine against our rate addiction and our addiction to material wealth might be to interfere in the way money flows. So besides technology, we need disciplines like psychology, philosophy, law, finance, economics, system dynamics and politics. And above all, we need dedicated and inspiring leadership that brings together all available information, sets up a time schedule and acts accordingly.

ANNEX A
PERIODIC TABLE OF ELEMENTS

(cut off after ^{103}Lr)

1a	2a	3a	4a	5a	6a	7a		8		1b	2b	3b	4b	5b	6b	7b	
^{1}H																	^{2}He
^{3}Li	^{4}Be											^{5}B	^{6}C	^{7}N	^{8}O	^{9}F	^{10}Ne
^{11}Na	^{12}Mg											^{13}Al	^{14}Si	^{15}P	^{16}S	^{17}Cl	^{18}Ar
^{19}K	^{20}Ca	^{21}Sc	^{22}Ti	^{23}V	^{24}Cr	^{25}Mn	^{26}Fe	^{27}Co	^{28}Ni	^{29}Cu	^{30}Zn	^{31}Ga	^{32}Ge	^{33}As	^{34}Se	^{35}Br	^{36}Kr
^{37}Rb	^{38}Sr	^{39}Y	^{40}Zr	^{41}Nb	^{42}Mo	^{43}Tc	^{44}Ru	^{45}Rh	^{46}Pd	^{47}Ag	^{48}Cd	^{49}In	^{50}Sn	^{51}Sb	^{52}Te	^{53}I	^{54}Xe
^{55}Cs	^{56}Ba	LA	^{72}Hf	^{73}Ta	^{74}W	^{75}Re	^{76}Os	^{77}Ir	^{78}Pt	^{79}Au	^{80}Hg	^{81}Tl	^{82}Pb	^{83}Bi	^{84}Po	^{85}At	^{86}Rn
^{87}Fr	^{88}Ra	AC															

LA	^{57}La	^{58}Ce	^{59}Pr	^{60}Nd	^{61}Pm	^{62}Sm	^{63}Eu	^{64}Gd	^{65}Tb	^{66}Dy	^{67}Ho	^{68}Er	^{69}Tm	^{70}Yb	^{71}Lu
AC	^{89}Ac	^{90}Th	^{91}Pa	^{92}U	^{93}Np	^{94}Pu	^{95}Am	^{96}Cm	^{97}Bk	^{98}Cf	^{99}Es	^{100}Fm	^{101}Md	^{102}No	^{103}Lr

ANNEX B
PRIMARY PRODUCTION AND RESERVES OF METALS

	Revised 2007 Production [T]	Reserves [T]	R/P-ratio at 2% prod.growth	
Ag	20800	270000	12	
Al	50500000	6750000000	65	production and reserves based on bauxite (factor 0.25)
As	55900	1118000	17	world reserves estimated at 20 times annual prod (USGS)
Au	2380	47000	17	
Ba	4578000	114000000	20	estimated fromBaSO4 (factor 0.6)
Be	180	18830	57	US reserves multiplied with world/US production factor
Bi	6300	320000	35	
Cd	20400	490000	20	
Co	65500	7100000	58	
Cr	5375000	2000000000	>70	production estimated from chromite ore (factor 0.25), reserves 1/6th of resources as with Cu,Fe
Cs	unknown	70000	unknown	cesium
Cu	15400000	550000000	27	
Fe	1000000000	75000000000	46	estimated from iron ore (factor 0.5), production corresponds with pig iron production
Ga	80	unknown	unknown	most Ga reserves in bauxite
Ge	100	unknown	unknown	reserves of US are 450T
Hg	1170	46000	29	
Hf	unknown	518500	unknown	estimated from HfO2 (factor 0.85)
In	563	unknown	unknown	most In reserves in Zn ores
Li	25800	4100000	>70	production excl US
Mg	5139000		>70	production is Mg compounds + metal, reserves virtually unlimited (also derived from seawater)

Mn	12600000	500000000	29	
Mo	205000	8600000	31	
Nb	60400	2700000	32	
Ni	1660000	70000000	31	
Pb	3770000	79000000	18	
PGM	432	71000	>70	Platinum Group Metals; reserves Pt,Pd,Rh,Ru,Ir,Os; production only Pt+Pd
REM	108000	76600000	>70	Rare Earth Metals; La + Y; estimated from RE2O3 (factor 0.87)
Re	51	2500	35	
Sb	170000	2100000	11	antimony
Sn	320000	5600000	15	
Sr	511000	6800000	12	
Ta	815	130000	>70	
Te	unknown	22000	unknown	production in Peru and Canada combined was 41 tons
Ti	3927000	438000000	59	estimated from TiO2 (factor 0.6), production added with 153000T sponge production
Tl	10	380	28	thallium
Y	7030	426600	40	estimated from Y2O3 (factor 0.79)
Zn	10900000	180000000	14	
Zr	1430000	37740000	21	reserves based on ZrO2 (factor 0.74)
V	58500	13000000	>70	
W	54500	3000000	37	

Based on "Mineral commodity summaries 2009" by U.S. Geological Survey

Printed in Dunstable, United Kingdom

91222430R00068